The book series "SDGs and Textiles" addresses the strategies to achieve sustainable development goals (SDGs) in the present, past, and future. It presents books about the present and future policies of textile ministries of different countries, and books related to sustainability education around different parts of the world in the textile sector. Moreover, it would welcome the conference proceeding related to SDGs and Textiles. The series would cover books comparing the sustainability and SDGs of different institutions and countries. The individual book volumes in the series are thematic. The goal of each book is to give readers a comprehensive overview of a different area of sustainability in the textile sector. As a collection, the series provides valuable resources to a broad audience in academia, the research community, industry, and anyone looking to expand their knowledge of SDGs and Textiles.

Textiles and life are together – life cannot be separated from textiles as it is the most important need for human beings after food. In 2015, the United Nations General Assembly proposed 17 interlinked global goals to be achieved by 2030. Since then, academia and industry have paid much attention to achieving these goals. Textile found its close relation with almost all of these 17 goals.

SDG 1 - No Poverty: Poverty would never be overcome by a charity only; it is essential to develop people's skills to have a better and wealthy life. Thus, the textile can be considered an excellent discipline to achieve this goal by creating jobs and small and medium businesses.

SDG 2 - Zero Hunger: Through the effective utilization of advanced application of Agrotech Textiles, it is possible to have higher crop yields and save crops from rough weather, unexpected rains, floods, insects, etc.; thus, geotextiles play an essential in achieving this goal of sustainable development.

SDG 3 - Good Health & well-being: There has been much health consciousness after Covid19, and medical textiles assist in getting good health and well-being.

SDG 4 - Learning & Education: Textile or fashion has remained a significant discipline for societies for ages, and there has always remained much to explore in this field. Textile-related universities may play a vital role by offering free access to their education resources, training and spreading information among the locals.

SDG 5 - Gender Equality: The textile sector is one of the industrial sectors that accepted gender equality long ago; in particular, the garment sector has more females than males. Thus, the textile sector has been doing gender equality. Moreover, there has been a recent trend for Gender Neutral Clothing, which need worth studying and may further assist gender equality.

SDG 6 - Clean Water & Sanitation: Textiles could be achieved through filtration, and of course, textile is one of the critical materials for filtration.

SDG 7 - Affordable & Clean Energy: With the recent advancement in material science and engineering, the textile sector has come on the front for, not only by using this clean energy during textile production but also by assisting the production of this clean energy, either in the form of wind turbines blades made of textile composites or by energy harvesting from T-Shirts, etc.

SDG 8 - Decent Work: Recently, there has been much attention that the textile workers are not paid well, labor rights are not cared about, etc.

SDG 9 - Industry and innovation: Textile Industry always follows innovation; the textile companies that do not chase innovation cannot survive in the market.

SDG 10 - Reduced Inequalities: Getting better life and well-being would help reduce inequalities in the textile industry.

SDG 11 - Sustainable Cities: Sustainable Textile Cities through Buildtech and transport textiles.

SDG 12 - Consumption and Production: Textile and garment consumption and production all come under.

SDG 13 - Climate Action: Oekotech or Ecotech Textile, waste management of textiles are upfront to achieve this goal of sustainable development.

SDG 14 - Life Below Water: Mitigating microfiber waste in rivers and oceans may come under the context of it. There has been much attention on this subject after passing the bill at the parliament level of the UK.

SDG 15 - Life on Land: Geotech or Geotextiles studies life on land.

SDG 16 - Peace, Justice, and Strong Institutions: Protective textiles are doing their best to achieve peace, justice, and strong institutions.

SDG 17 - Partnerships for the Goals: The application of textiles to achieve sustainable development goals is only an example. In all textiles sectors, combined efforts of all the goals are essential to achieve true sustainability.

Aníbal Enrique Toscano-Hernández ·
Saúl Alfonso Esparza-Rodríguez ·
Gabino García-Tapia
Editors

Sustainability, Stakeholders and Marketing in the Textile Sector

Conceptual and Practical Cases for the Americas

Springer

Editors
Aníbal Enrique Toscano-Hernández
Universidad del Sinú
Montería, Córdoba, Colombia

Gabino García-Tapia
Facultad de Contaduría y Administración
Universidad Nacional Autónoma de México
Mexico City, Mexico

Saúl Alfonso Esparza-Rodríguez
Facultad de Contaduría y Administración
Universidad Nacional Autónoma de México
Mexico City, Mexico

ISSN 2948-1236 ISSN 2948-1244 (electronic)
SDGs and Textiles
ISBN 978-981-97-6951-3 ISBN 978-981-97-6952-0 (eBook)
https://doi.org/10.1007/978-981-97-6952-0

This Springer imprint is published by the registered company Springer Nature Singapore Pte Ltd.
The registered company address is: 152 Beach Road, #21-01/04 Gateway East, Singapore 189721, Singapore

Preface

Sustainability in the textile industry has emerged as a topic of vital importance, both in academic research and in practical application. This book, entitled *Sustainability, Stakeholders and Marketing in the Textile Sector: Conceptual and Practical Cases for the Americas*, responds to the imperative need to understand and improve sustainable practices in a crucial sector of the global economy. The textile industry is not only a significant economic driver but also a considerable source of environmental and social impacts.

The relevance of this work is particularly notable in the context of the United Nations' 2030 Sustainable Development Goals (SDGs), which serve as a guiding framework for the discussions and analyses presented. The objective of this work is to provide a comprehensive and multidisciplinary vision of the integration of sustainability principles in textile value chains in the Americas.

The editors, Dr. Aníbal Enrique Toscano Hernández, Dr. Saúl Alfonso Esparza-Rodríguez, and Dr. Gabino García-Tapia have brought together a group of leading researchers and professionals in the sector to offer case studies and theoretical analyses that address various dimensions of sustainability in the textile industry. The structure of the book is divided into three main sections: "Section I. State of the Art", "Section II. Management sustainable practices in textile industry", and "Section III. Marketing, consumers, and fair trade in the textile industry supply chain". Through carefully crafted chapters, readers will find a combination of qualitative and quantitative methodologies that provide valuable and actionable insights. This multidisciplinary and multi-methodological approach not only enriches theoretical understanding but also offers practical recommendations for the implementation of sustainable strategies.

We hope that this book will serve as a source of knowledge and inspiration for academics, industry professionals, and policymakers seeking to promote more sustainable practices in the textile industry. By taking a critical and thoughtful perspective, this work not only documents existing advances but also stimulates new research and actions that contribute to a more sustainable future for the textile sector and beyond.

Montería, Colombia — Aníbal Enrique Toscano-Hernández
Mexico City, Mexico — Saúl Alfonso Esparza-Rodríguez
Mexico City, Mexico — Gabino García-Tapia

Introduction

The textile industry is one of the most dynamic and complex sectors of the global economy, characterized by an extensive value chain that ranges from the production of fibers to the marketing of final products. This sector faces numerous challenges related to sustainability, including high consumption of natural resources, intensive use of chemicals, and significant generation of waste and greenhouse gas emissions. These factors not only affect the environment but also pose social and economic risks.

This book, *Sustainability, Stakeholders and Marketing in the Textile Sector: Conceptual and Practical Cases for the Americas* examines how these challenges are being addressed in the Americas. The region, which includes emerging and developed economies, presents a diversity of economic, social, and environmental contexts that offer fertile ground for sustainability research. Through a series of case studies and theoretical analyses, this work explores sustainable practices in the textile industry and their alignment with the Sustainable Development Goals (SDGs). The structure of the book is organized into three main sections.

Initially, Part I provides an overview of the current state of sustainability in the textile industry, exploring the global trends, challenges, and opportunities facing the sector (Chaps. 1 and 2), highlighting recent studies and technological advances that are driving sustainability, aligning with SDGs 9 (Industry, Innovation, and Infrastructure) and 12 (Responsible Consumption and Production).

Next, Part II presents case studies that examine sustainable practices in the management of the textile industry, including analyses on eco-innovation in exporting companies (Chap. 3), transformational leadership in textile companies (Chap. 4), and obstacles to innovation in textile SMEs (Chap. 5), connecting with SDGs 8 (Decent Work and Economic Growth) and 13 (Climate Action). In addition, the implementation of reverse logistics and the circular economy as fundamental tools for sustainability is discussed (Chap. 6), linked to SDGs 11 (Sustainable Cities and Communities) and 12.

Finally, Part III focuses on green marketing and fair trade in the textile supply chain, presenting case studies on marketing strategies in sustainable fashion companies (Chap. 7), the green swan theory and its relationship to regenerative capitalism

(Chap. 9), as well as the influence of these strategies on consumer perception and behavior (Chap. 8), aligning with SDGs 10 (Reduced Inequalities) and 12.

Together, the chapters in this book provide a comprehensive overview of the various strategies and practices that are being implemented in the textile sector to improve its sustainability. In doing so, we hope to contribute to the global dialogue on sustainability in the textile industry and encourage the adoption of more responsible and sustainable practices across the value chain.

Contents

Chapter 1
Geospatial Analysis of the Sustainable Value Chain in the Textile Sector of the United States

Gabino García-Tapia, Sanjay Nanwani, and José Fernando Acosta López

Introduction

In the recent years, the use of products in the textile industry that represent an important environmental impact has drawn increasing concerns in the global society, considering that widely used products such as conventional PET contributes to pollution across its entire value chain, including manufacturing, use, and disposal, exacerbating resource depletion in the process, with adverse effects on land, water, and air quality (Palacios-Mateo et al. 2021); also, there is an increasing public concern over rising water consumption, related water scarcity issues, and corporate water accounting across entire supply chains as well (Joa et al. 2014).

In that sense, one remarkable example regards the fast fashion industry, which has significantly increased global production and consumption, that represent shorter lifespans and thus, more pollution to the environment, where the extent of the impacts remains highly related to key environmental indicators such as energy consumption, climate and water resource impacts, including also social indicators such as wages and employment (Peters et al. 2021); that situation is bringing international strategies for circular fashion as a result, aiming to minimize the environmental impact of the industry by creating a more sustainable system, based on relevant stakeholder engagement including international bodies, non-governmental organizations (NGOs), and sector-specific organizations (Manshoven and Van Opstal 2022).

G. García-Tapia (✉)
Faculty of Accountant and Administration, National Autonomous University of Mexico, Coyoacán, Mexico
e-mail: gabino@unam.mx

S. Nanwani
Department of Liberal Arts and Communication, Arkansas State University Campus Queretaro, Querétaro, Mexico

J. F. A. López
University of Sinú, Montería, Colombia

A. E. Toscano-Hernández et al. (eds.), *Sustainability, Stakeholders and Marketing in the Textile Sector*, SDGs and Textiles, https://doi.org/10.1007/978-981-97-6952-0_1

Such situation is highly correlated to international markets and even geopolitics, since most of the biggest economies are interconnected through direct and intermediate trade, with important statistics among a diverse number of industries, international value chains are supported by economic and political links that includes imports and exports of machinery, rubber and plastics, and textiles (Gnidchenko 2023); in that sense, designing, operating and managing global value chains, was disrupted by the outbreak of COVID-19, bringing a global situation that spanned environmental impacts from production to consumption, changing important relationships for the United States of America (USA) with China and India, mainly in relation to textiles, wearing apparel, leather, and related products (Bagaria 2022; Bagchi and Bhattacharyya 2021; Chowdhury et al. 2023; Köhler et al. 2008); such environmental impacts where dramatically increased specially in personal protective equipment (PPE) for single-use (basically, constructed from polymer plastic-based synthetic fibers), showing that the COVID-19 pandemic represented a widespread surge in the use of disposable medical devices, exacerbating plastic and microplastic pollution (Pellegrini et al. 2023); consequently, in the present days, there is an urgent need for in-depth to identify opportunities for high-end upgrading in these sectors, mostly by industrial upgrading across technology, organization, and environmental dimensions (Zeying 2014).

Now, given the evident importance of supply chains for competitiveness and positive performance of industries, adopting both lean and green paradigms is crucial to create the conditions for sustaining competitive advantage in the long term (Thanki and Thakkar 2018); Since there are some other relevant sustainable alternatives that bring opportunities in terms of product development, improvement and fostering innovation, replacing plastics and shifting production trends that can even stimulate engagement for local value chains and preserve local heritage in communities (Løvbak Berg et al. 2023); this also represented an advancement in the development of traditional manufacturing sectors, including the textile and apparel industries, by technological progress and technical efficiency (Pellegrini et al. 2023).

One innovative approach that contributes to create the conditions for a company to adopt sustainable practices is upcycling, which is a process that involves repairing, reusing, repurposing and refurbishing, to improve the quality and lifespans of materials and products while reducing waste, and promoting sustainable consumer engagement, focused on creating opportunities for collaboration involving a diverse range of interest parties (Singh et al. 2019).

From decades, the textile industry has represented one of the highest GHG emitters in the planet, according to the diversity of processes and products related to the manufacture of fibers, textiles and apparel products, that require significant use of fossil fuel, accounting for nearly 10 percent of the total carbon impact (TTRI Taiwan Textile Research Institute 2013).

In such reality, the textile and garment industry contributes to a significant carbon emissions, ranging from 6 to 8% approximately of global carbon emissions (Imran et al. 2023), becoming one of the most polluting industries, representing an ecological footprint caused by the enormous amount of energy, water and chemical use, representing around 8–10% of global carbon emissions (Filho et al. 2022), where the textile value chain is estimated to be responsible for between 2 and 8% of global greenhouse gas (GHG) emissions, as well as significant pollution, water extraction and biodiversity impacts (United Nations Environment Programme 2023).

In such context, the present research works aim to analyze the sustainable value chain in the American textile industry, identifying the most relevant papers in the subject, as well as identification of the geospatial indicators of textile industry-related information, which contributes to the understanding regarding the need for new public policies addressing such environmental for the American textile industry, that would contribute to the development of a sustainable economy, and enhancing a sustained competitive advantage for such industry in a competitive global market, where the research goals will contribute to understand the following research question:

- *What is the level of regional concentration of the textile value chain industry in terms of GHG emissions per state in the United States of America?*

Literature Review

The relationships between textile industry and sustainability can be traced globally, where the relevance of such productive activities is relevant for economic systems around the world; since textile represents a relevant industrial sector, the studies related to its evident environmental and social impact are especially important in developing countries, where this represents a known issue, concerning both environmental and economic-social facets of sustainability (Alves et al. 2023); it is known that, in the case of the fashion sector, clothing manufacturing and transportation produce a large amount of greenhouse gas emissions and thus, by analyzing the effects of the environmental regulation on industrial and efficiency within the textile sector, the findings will contribute with theoretical significance to promote environmentally friendly and sustainable progress (Ju et al. 2023), which is correspondent to the raising of environmental and sustainable traceability as a critical priority toward sustainable and circular value chains in the fashion industry (Joséphine et al. 2022).

For what it comes to the progress done in the transition toward sustainable production and consumption patterns, subjects such as transparency and traceability along the supply chain are fundamental for transition, including creation of multi-stakeholders' clusters, training programs and skills development along the textile product life cycle stages (Luján-Ornelas et al. 2020); this is especially relevant since the textile industry is continuously drawing attention to its negative sustainability impacts along their value chains (Olofsson and Mark-Herbert 2020).

In response to the challenges posed by the environmental impact of textile industries, promoting inclusive and sustainable development must also include employment and manufacturing value added, as well as export strategies, actively engaging the promotion of sustainable and inclusive growth at the centre of its action, emphasizing on economic and social inclusiveness and active involvement of vulnerable groups in the society (Gallico 2020).

Nowadays, sustainable manufacturing is gaining prominence by procuring assessment of the current state of sustainability in the related industries, identifying potential sustainable manufacturing practices in the domains of any given manufacturing firm's value chain (Pande and Adil 2022); also, sustainable related aspects and their verification are becoming increasingly associated with the comprehensive sustainability agenda from textile companies (gate-to-gate), as well as the upstream stages of the value chain (cradle-to-gate), based on process models and textile production process (Stellmach et al. 2022).

Considering the European Union, the Textile Strategy for Sustainable and Circular Textiles represents a relevant opportunity to engage large economies in promoting circularity along the value chain of textiles, where German manufacturing textile industries apply a circular business model strategies including upcycling and recycling, product-as-a-service, repair and reuse, and material innovation (Arnold et al. 2023).

In the case of Bangladesh, the textile wastewater management in the production value chain shows the degree of adverse effects of textile wastewater on freshwater resources, including the obstruction of socio-economic development and environmental sustainability, highlighting important distinctions between compliant and non-compliant textile industries, indicating challenges that hinder the sustainability of textile effluent treatment as well, by evidencing noticeable absences of techno-ecology, pro-environmental behaviors, stringent regulations, reliable surveillance, and collaborative actions (Gomes et al. 2024).

Given the context and state of the knowledge in the subject, this research aims to provide an understanding of the sustainable value chain of the textile industry in the USA, using a statistical, descriptive, and quantitative approach to the main geospatial indicators regarding the textile industry.

Methods

The present work includes a geospatial analysis of the triple bottom line of the sustainable value chain textile industry using descriptive and graphical quantitative analysis techniques per American state, considering a multivariable analysis per regions, including environmental impacts, using data from the Environmental Protection Agency (EPA), in relation to Green House Gases Emissions reported by facilities located in every state of USA, with data available as of August 18, 2023, reporting of greenhouse gas (GHG) data, where approximately 8,000 facilities are required to report their emissions annually; as well as economic factors and social responsibility practices, available in the environmental agenda from the current government of the USA,[1] and in the FLIGHT digital platform, which includes information about GHG emission from such nation.

[1] https://www.whitehouse.gov/climate/#:~:text=Reducing%20U.S.%20greenhouse%20gas%20emissions,clean%20energy%20to%20disadvantaged%20communities.

Results

The analysis incorporates data sourced from FLIGHT, a digital platform designed to offer comprehensive information on greenhouse gas (GHG) emissions from major facilities across the country. FLIGHT is an integral component of the Greenhouse Gas Reporting Program (GHGRP), providing invaluable insights into emissions data. It's crucial to note that all emissions data is reported in metric tons of carbon dioxide equivalent, utilizing Global Warming Potentials (GWPs) from the IPCC's Fourth Assessment Report (AR4) as of August 18th, 2023. The information regards direct-emitting facilities report emissions from each source category, and these emissions can generally be categorized as either combustion or process emissions (emissions from fuel combustion comprise carbon dioxide (CO_2), methane (CH_4), and nitrous oxide (N_2O) emitted from combustion of a fossil fuel (e.g., coal, natural gas, petroleum products) or biomass feedstock (e.g., wood, landfill gas)).[2] This encompasses data from all American states, considering GHGs such as CH_4, and includes information from all emitters, which is detailed in the following Table 1.1.

Table 1.1 GHG emissions in units of metric tons of carbon dioxide equivalent using GWP's from IPCC's AR4, as of 2023 per every state and parent companies in the United States of America (USA)

City name	County name	State	Parent companies	GHG quantity (metric tons CO_2e)	Subparts*
DECATUR	MORGAN COUNTY	AL	3 M CO (100%)	17	C, L, OO
CORDOVA	ROCK ISLAND COUNTY	IL	3 M CO (100%)	18	C, L, OO
BARCELONETA	BARCELONETA MUNICIPIO	PR	ABBVIE INC (100%)	33	C
Coushatta	RED RIVER PARISH	LA	ADA CARBON SOLUTIONS LLC (100%)	1	C
Southport	BRUNSWICK COUNTY	NC	ARCHER DANIELS MIDLAND CO (100%)	42	C
HOPEWELL	HOPEWELL CITY	VA	ADVANSIX INC (100%)	254	C, G, PP
LUVERNE	ROCK COUNTY	MN	GEVO INC (100%)	1	C, II
BORGER	HUTCHINSON COUNTY	TX	AGRIUM US INC (100%)	235	C, G, PP

(continued)

[2] https://www.epa.gov/sites/default/files/2017-12/documents/ghgrp_methodology_factsheet.pdf.

Table 1.1 (continued)

City name	County name	State	Parent companies	GHG quantity (metric tons CO_2e)	Subparts*
KENNEWICK	BENTON COUNTY	WA	AGRIUM US INC (100%)	1	C, U, V
PORT NECHES	JEFFERSON COUNTY	TX	AMERICAN AIR LIQUIDE HOLDINGS INC (100%)	2	C
GEISMAR	ASCENSION PARISH	LA	AMERICAN AIR LIQUIDE HOLDINGS INC (100%)	210	C
BAYTOWN	HARRIS COUNTY	TX	AIR PRODUCTS & CHEMICALS INC (100%)	28	C
WILMINGTON	LOS ANGELES COUNTY	CA	AIR PRODUCTS & CHEMICALS INC (100%)	174	P
PASADENA	HARRIS COUNTY	TX	ALBEMARLE CORP (100%)	22	C
MAGNOLIA	COLUMBIA COUNTY	AR	ALBEMARLE CORP (100%)	39	C
Dunbar	KANAWHA COUNTY	WV	ALTIVIA SERVICES LLC (100%)	50	C
HAVERHILL	SCIOTO COUNTY	OH	ALTIVIA CORP (100%)	13	C
PASADENA	HARRIS COUNTY	TX	ARKEMA DELAWARE INC (50%); NIPPON SHOKUBAI AMERICA INDUSTRIES INC (50%)	35	C
ST. JAMES	ST. JAMES PARISH	LA	CHEVRON CORP (50%); TRINSEO PLC (50%)	426	C
JUNCOS	JUNCOS MUNICIPIO	PR	AMGEN INC (100%)	37	C
STERLINGTON	OUACHITA PARISH	LA	ARUBA INVESTMENTS INC (100%)	48	C, V
St. David	COCHISE COUNTY	AZ	FREEPORT-MCMORAN INC (50%); SOUTHWEST ENERGY LLC (50%)	13	C, V
BEAUMONT	JEFFERSON COUNTY	TX	ARKEMA DELAWARE INC (100%)	26	C
PASADENA	HARRIS COUNTY	TX	ARKEMA DELAWARE INC (100%)	42	C

(continued)

Table 1.1 (continued)

City name	County name	State	Parent companies	GHG quantity (metric tons CO_2e)	Subparts*
CALVERT CITY	MARSHALL COUNTY	KY	ARKEMA DELAWARE INC (100%)	29	C, L, OO
ALVIN	BRAZORIA COUNTY	TX	ASCEND PERFORMANCE MATERIALS HOLDINGS INC (100%)	123	C, X
PHILADELPHIA	Philadelphia	PA	ADVANSIX INC (100%)	74	C
La Porte	HARRIS COUNTY	TX	AIR PRODUCTS & CHEMICALS INC (100%)	7	C
Port Arthur	JEFFERSON COUNTY	TX	AIR PRODUCTS & CHEMICALS INC (100%)	457	C, P, PP
Pasadena	HARRIS COUNTY	TX	ALBEMARLE CORP (100%)	51	C
FORT WORTH	TARRANT COUNTY	TX	ALCON LABORATORIES INC (100%)	8	C
Oregon	LUCAS COUNTY	OH	ALPONT LLC (100%)	22	C, X
DECATUR	Morgan	AL	ASCEND PERFORMANCE MATERIALS HOLDINGS INC (100%)	374	C, P, PP
Cantonment	ESCAMBIA COUNTY	FL	ASCEND PERFORMANCE MATERIALS HOLDINGS INC (100%)	344	C, E, P, TT, V
IRVINE	ORANGE COUNTY	CA	B BRAUN MEDICAL US (100%)	24	C
BEAUMONT	JEFFERSON COUNTY	TX	BASF CORP (100%)	32	C
GEISMAR	ASCENSION PARISH	LA	BASF CORP (100%)	1130	C, X
PASADENA	HARRIS COUNTY	TX	BASF CORP (100%)	18	C
MC INTOSH	WASHINGTON COUNTY	AL	BASF CORP (100%)	35	C
PALMYRA	MARION COUNTY	MO	BASF CORP (100%)	57	C
Wyandotte	WAYNE COUNTY	MI	BASF CORP (100%)	17	C

(continued)

Table 1.1 (continued)

City name	County name	State	Parent companies	GHG quantity (metric tons CO_2e)	Subparts*
FREEPORT	BRAZORIA COUNTY	TX	BASF CORP (100%)	280	C
PORT ARTHUR	JEFFERSON COUNTY	TX	BASF CORP (100%)	10392	C, X
KANSAS CITY	JACKSON COUNTY	MO	BAYER US HOLDING LP (100%)	24	C
CAMBRIDGE	MIDDLESEX COUNTY	MA	BIOGEN INC (100%)	14	C
ULYSSES	GRANT COUNTY	KS	BIRLA CARBON USA INC (100%)	6	C, X
NORTH BRUNSWICK	MIDDLESEX COUNTY	NJ	BRISTOL-MYERS SQUIBB CO (100%)	26	C
MUSCATINE	MUSCATINE COUNTY	IA	BAYER US HOLDING LP (100%)	36	C
LULING	ST. CHARLES PARISH	LA	BAYER US HOLDING LP (100%)	40	C
PASADENA	Harris	TX	AMERICAN AIR LIQUIDE HOLDINGS INC (100%)	2515	C
Port Arthur	JEFFERSON COUNTY	TX	BAYPORT POLYMERS LLC (100%)	59774	C, X
Belpre	WASHINGTON COUNTY	OH	ORION ENGINEERED CARBONS LLC (100%)	4	C, X
BIG SPRING	HOWARD COUNTY	TX	TOKAI CARBON CB LTD (100%)	5	C, X
Borger	HUTCHINSON COUNTY	TX	ORION ENGINEERED CARBONS LLC (100%)	26	C, X
Borger	HUTCHINSON COUNTY	TX	TOKAI CARBON CB LTD (100%)	22	C, X
HENDERSON	CLARK COUNTY	NV	EMD ACQUISITION LLC (100%)	13	C
FRANKLIN	ST. MARY PARISH	LA	CABOT CORP (100%)	9	C, X
VILLE PLATTE	EVANGELINE PARISH	LA	CABOT CORP (100%)	5	C, X
CATLETTSBURG	BOYD COUNTY	KY	KURARAY HOLDINGS USA INC (100%)	22	C
WINDSOR	WELD COUNTY	CO	CARESTREAM HEALTH INC (100%)	15	C

(continued)

Table 1.1 (continued)

City name	County name	State	Parent companies	GHG quantity (metric tons CO_2e)	Subparts*
DONALDSONVILLE	ASCENSION PARISH	LA	CF INDUSTRIES HOLDINGS INC (100%)	1382	C, G, V
YAZOO	YAZOO COUNTY	MS	CF INDUSTRIES HOLDINGS INC (100%)	224	C, G, V
SERGEANT BLUFF	WOODBURY COUNTY	IA	CF INDUSTRIES HOLDINGS INC (100%)	485	C, G, PP, V
CHANNELVIEW	HARRIS COUNTY	TX	LYONDELLBASELL INDUSTRIES INC (100%)	4988	C, MM, X
DEEPWATER	SALEM COUNTY	NJ	THE CHEMOURS CO (100%)	1	C, L, OO, TT
FAYETTEVILLE	BLADEN COUNTY	NC	THE CHEMOURS CO (100%)	14	C, L, OO
PASS CHRISTIAN	HARRISON COUNTY	MS	THE CHEMOURS CO (100%)	160	C, EE
EL DORADO	UNION COUNTY	AR	THE CHEMOURS CO (100%)	1	C, L, OO
LOUISVILLE	JEFFERSON COUNTY	KY	THE CHEMOURS CO (100%)	13	C, O, TT
WASHINGTON	WOOD COUNTY	WV	THE CHEMOURS CO (100%)	102	C, L, O, OO
CHEROKEE	COLBERT COUNTY	AL	LSB INDUSTRIES INC (100%)	58	C, G, PP, V
Sweeny	BRAZORIA COUNTY	TX	CHEVRON CORP (50%); PHILLIPS 66 (50%)	5905	C, X
PORT ARTHUR	JEFFERSON COUNTY	TX	CHEVRON CORP (50%); PHILLIPS 66 (50%)	3737	C, X
PASADENA	HARRIS COUNTY	TX	CELANESE CORP (100%)	262	C, P, X
COFFEYVILLE	MONTGOMERY COUNTY	KS	CVR ENERGY INC (100%)	23	C, G, PP, V
CENTERVILLE	ST. MARY PARISH	LA	BIRLA CARBON USA INC (100%)	6	C, X
PHENIX CITY	RUSSELL COUNTY	AL	CSRC USA CORP (100%)	8	C, X
PONCA CITY	KAY COUNTY	OK	CSRC USA CORP (100%)	32	C, X

(continued)

Table 1.1 (continued)

City name	County name	State	Parent companies	GHG quantity (metric tons CO_2e)	Subparts*
SUNRAY	MOORE COUNTY	TX	CSRC USA CORP (100%)	9	C, X
Proctor	WETZEL COUNTY	WV	COVESTRO LCC (100%)	14	C
MEMPHIS	SHELBY COUNTY	TN	COVORO MINING SOLUTIONS LLC (100%)	14	C
BRADLEY	KANKAKEE COUNTY	IL	CSL BEHRING LLC (100%)	24	C
WILLOW ISLAND	PLEASANTS COUNTY	WV	CYTEC INDUSTRIES INC (100%)	19	C
Gregory	SAN PATRICIO COUNTY	TX	THE CHEMOURS CO (100%)	5	C, L, OO
BELLE	Kanawha	WV	THE CHEMOURS CO (100%)	49	C, TT
NEW JOHNSONVILLE	Humphreys	TN	THE CHEMOURS CO (100%)	44	C, EE
Oregon	LUCAS COUNTY	OH	CHEMTRADE LOGISTICS INC (100%)	14	C
Belle Chasse	PLAQUEMINES PARISH	LA	CHEVRON CORP (100%)	68	C
Baytown	HARRIS COUNTY	TX	CHEVRON CORP (50%); PHILLIPS 66 (50%)	14996	C, X
Cheyenne	LARAMIE COUNTY	WY	DYNO NOBEL HOLDINGS USA INC (100%)	79	C, G, PP, V
Cheyenne	LARAMIE COUNTY	WY	HF SINCLAIR CORP (100%)	68	C
Louisville	JEFFERSON COUNTY	KY	CLARIANT CORP (100%)	12	C
El Dorado	UNION COUNTY	AR	CLEAN HARBORS INC (100%)	17	C, O, OO
Newport	NEW CASTLE COUNTY	DE	SUN CHEMICAL (100%)	15	C
Waggaman	JEFFERSON PARISH	LA	CSTN HOLDINGS INC (100%)	125	C, X
HARBOR BEACH	HURON COUNTY	MI	CORTEVA INC (100%)	29	C

(continued)

Table 1.1 (continued)

City name	County name	State	Parent companies	GHG quantity (metric tons CO_2e)	Subparts*
PITTSBURG	CONTRA COSTA COUNTY	CA	CORTEVA INC (100%)	14	C
Carville	IBERVILLE PARISH	LA	SABIC INNOVATIVE PLASTICS US LLC (50%); TOTAL HOLDINGS USA INC (50%)	152	C
New Castle	NEW CASTLE COUNTY	DE	CRODA INTERNATIONAL PUBLIC LC (100%)	45	C
DECATUR	MORGAN COUNTY	AL	DAIKIN HOLDINGS AMERICA INC (100%)	22	C, L, O, OO
CARROLLTON	CARROLL COUNTY	KY	DOW INC (100%)	78	C, TT
FREEPORT	BRAZORIA COUNTY	TX	DOW INC (100%)	3229	C, DD, TT, W, X
BELVIDERE	WARREN COUNTY	NJ	DSM HOLDING CO USA INC (100%)	21	C
LA PLACE	ST. JOHN THE BAPTIST PARISH	LA	DUPONT DE NEMOURS INC (100%)	48	C
LOUISIANA	PIKE COUNTY	MO	DYNO NOBEL HOLDINGS USA INC (100%)	3	C, V
BATON ROUGE	EAST BATON ROUGE PARISH	LA	Deltech Monomers Bidco Holdings, LLC (100%)	46	C
Norco		LA	DARLING INGREDIENTS INC (50%); VALERO ENERGY CORP (50%)	19	C
Pasadena	HARRIS COUNTY	TX	GLENCOE CAPITAL (100%)	23	C
Midland	MIDLAND COUNTY	MI	DOW INC (100%)	18	C
NEWARK	NEW CASTLE COUNTY	DE	INTERNATIONAL FLAVORS & FRAGRANCES INC (100%)	13	C
Deer Island	COLUMBIA COUNTY	OR	DYNO NOBEL HOLDINGS USA INC (100%)	33	C, G, PP, V

(continued)

Table 1.1 (continued)

City name	County name	State	Parent companies	GHG quantity (metric tons CO_2e)	Subparts*
Waggaman	JEFFERSON PARISH	LA	DYNO NOBEL HOLDINGS USA INC (100%)	89	C, G, PP
LAWRENCEVILLE	MERCER COUNTY	NJ	BRISTOL-MYERS SQUIBB CO (100%)	18	C
CARSON	LOS ANGELES COUNTY	CA	ECOVYST INC (100%)	17	C
EL DORADO	UNION COUNTY	AR	LSB INDUSTRIES INC (100%)	154	C, G, PP, V
CASTLE HAYNE	NEW HANOVER COUNTY	NC	ELEMENTIS GLOBAL LLC (100%)	58	C, U
KALAMA	COWLITZ COUNTY	WA	LANXESS CORP (100%)	31	C
LA PORTE	HARRIS COUNTY	TX	LYONDELLBASELL INDUSTRIES INC (100%)	3856	C, X
PASADENA	HARRIS COUNTY	TX	LYONDELLBASELL INDUSTRIES INC (100%)	8	C, PP, X
CLINTON	CLINTON COUNTY	IA	LYONDELLBASELL INDUSTRIES INC (100%)	2686	C, X
MORRIS	GRUNDY COUNTY	IL	LYONDELLBASELL INDUSTRIES INC (100%)	1279	C, X
CORPUS CHRISTI	NUECES COUNTY	TX	LYONDELLBASELL INDUSTRIES INC (100%)	8527	C, X
NEW JOHNSONVILLE	HUMPHREYS COUNTY	TN	ERACHEM COMILOG INC (100%)	33	C
Nekoosa	WOOD COUNTY	WI	ERCO (US) HOLDINGS INC (100%)	12	C
MAPLETON	PEORIA COUNTY	IL	EVONIK CORP (100%)	30	C, P
THEODORE	MOBILE COUNTY	AL	EVONIK CORP (100%)	63	C
CHESTER	DELAWARE COUNTY	PA	EVONIK CORP (100%)	11	C
ETOWAH	MCMINN COUNTY	TN	EVONIK CORP (100%)	24	C

(continued)

Table 1.1 (continued)

City name	County name	State	Parent companies	GHG quantity (metric tons CO_2e)	Subparts*
LAFAYETTE	TIPPECANOE COUNTY	IN	EVONIK CORP (100%)	31	C
GREENSBORO	GUILFORD COUNTY	NC	EVONIK CORP (100%)	16	C
GARYVILLE	ST. JOHN THE BAPTIST PARISH	LA	EVONIK CORP (100%)	19	C
BATON ROUGE	EAST BATON ROUGE PARISH	LA	EXXON MOBIL CORP (100%)	2783	C, MM, X, Y
BAYTOWN	HARRIS COUNTY	TX	EXXON MOBIL CORP (100%)	4129	C, MM, X, Y
Proctor	Marshall	WV	WESTLAKE CHEMICAL CORP (100%)	241	C
EAST DUBUQUE	JO DAVIESS COUNTY	IL	CVR ENERGY INC (100%)	120	C, G, PP, V
LONGVIEW	Harrison	TX	Eastman Chemical Co (100%)	9418	C, D, TT, X
KINGSPORT	Sullivan	TN	Eastman Chemical Co (100%)	4815	C, TT, X
Texas City	GALVESTON COUNTY	TX	EASTMAN CHEMICAL CO (100%)	22	C
BATON ROUGE	EAST BATON ROUGE PARISH	LA	ECOVYST INC (100%)	20	C
CLINTON	VERMILLION COUNTY	IN	ELANCO US INC (100%)	32	C
CINCINNATI	Hamilton	OH	EMERY OLEOCHEMICALS LLC (100%)	83	C
BEAUMONT	Jefferson	TX	EXXON MOBIL CORP (100%)	4271	C, D, MM, X, Y
BATON ROUGE	EAST BATON ROUGE PARISH	LA	FORMOSA PLASTICS CORP USA (100%)	141	C, X
POINT COMFORT	CALHOUN COUNTY	TX	FORMOSA PLASTICS CORP USA (100%)	13547	C, X
OGDEN	WEBER COUNTY	UT	FRESENIUS MEDICAL CARE HOLDINGS INC (100%)	32	C
Channahon	Will	IL	INEOS USA LLC (100%)	55	C

(continued)

Table 1.1 (continued)

City name	County name	State	Parent companies	GHG quantity (metric tons CO_2e)	Subparts*
Greenwood	GREENWOOD COUNTY	SC	FUJIFILM HOLDINGS AMERICA CORP (100%)	9	C
Batesville	INDEPENDENCE COUNTY	AR	FUTUREFUEL CORP (100%)	249	C
SOUTH SAN FRANCISCO	SAN MATEO COUNTY	CA	ROCHE HOLDINGS INC (100%)	14	C
PASADENA	HARRIS COUNTY	TX	THE GOODYEAR TIRE & RUBBER CO (100%)	29	C
BALTIMORE	BALTIMORE CITY	MD	WR GRACE & CO (100%)	54	C
SULPHUR	CALCASIEU PARISH	LA	WR GRACE & CO (100%)	71	C
PORT LAVACA	CALHOUN COUNTY	TX	INEOS USA LLC (100%)	66	C, X
CRESTON	UNION COUNTY	IA	GREEN VALLEY CHEMICAL CORP (100%)	1	C, G
ZEBULON	WAKE COUNTY	NC	GLAXOSMITHKLINE SL HOLDINGS INC (100%)	4	C, OO
Gregory	SAN PATRICIO COUNTY	TX	EXXON MOBIL CORP (50%); SABIC US HOLDINGS LP (50%)	6029	C, X
CHICKASAW	MOBILE COUNTY	AL	HONEYWELL INTERNATIONAL INC (100%)	23	C
METROPOLIS	MASSAC COUNTY	IL	HONEYWELL INTERNATIONAL INC (100%)	2	C
BATON ROUGE	EAST BATON ROUGE PARISH	LA	HONEYWELL INTERNATIONAL INC (100%)	14	C, L, OO
GEISMAR	ASCENSION PARISH	LA	HONEYWELL INTERNATIONAL INC (100%)	50	C, L, O, OO
CONROE	MONTGOMERY COUNTY	TX	HUNTSMAN PETROCHEMICAL LLC (100%)	24	C
LAWRENCE	DOUGLAS COUNTY	KS	ICL SPECIALTY PRODUCTS NORTH AMERICA INC (100%)	20	C, U

(continued)

Table 1.1 (continued)

City name	County name	State	Parent companies	GHG quantity (metric tons CO_2e)	Subparts*
Gallipolis Ferry	MASON COUNTY	WV	ICL Industrial Products (100%)	11	C
JACKSONVILLE	DUVAL COUNTY	FL	INTERNATIONAL FLAVORS & FRAGRANCES INC (100%)	23	C
PASADENA	HARRIS COUNTY	TX	INEOS USA LLC (100%)	26	C
ALVIN	BRAZORIA COUNTY	TX	INEOS USA LLC (100%)	8936	C, X
Lima	ALLEN COUNTY	OH	INEOS USA LLC (100%)	19	C, X
TEXAS CITY	GALVESTON COUNTY	TX	INEOS AMERICAS LLC (100%)	152	C
ASHTABULA	ASHTABULA COUNTY	OH	INEOS PIGMENTS USA INC (100%)	16	C, EE
Ashtabula	ASHTABULA COUNTY	OH	INEOS PIGMENTS USA INC (100%)	95	C, EE
Texas City		TX	INEOS US CHEMICALS CO (100%)	122	C
Wando	Berkeley	SC	INEOS US CHEMICALS CO (100%)	51	C
NORTH CHARLESTON	CHARLESTON COUNTY	SC	INGEVITY CORP (100%)	11	C
DERIDDER	BEAUREGARD PARISH	LA	INGEVITY CORP (100%)	27	C
WICKLIFFE	BALLARD COUNTY	KY	INGEVITY CORP (100%)	43	C
VICTORIA	VICTORIA COUNTY	TX	KOCH INDUSTRIES INC (100%)	297	C, E, V
HOUSTON	HARRIS COUNTY	TX	KOCH INDUSTRIES INC (100%)	692	C
ORANGE	ORANGE COUNTY	TX	KOCH INDUSTRIES INC (100%)	75	C
CALVERT CITY	MARSHALL COUNTY	KY	ASHLAND GLOBAL HOLDINGS INC (100%)	34	C,TT
LIMA	ALLEN COUNTY	OH	ASHLAND GLOBAL HOLDINGS INC (100%)	62	C
TEXAS CITY	GALVESTON COUNTY	TX	ASHLAND GLOBAL HOLDINGS INC (100%)	18	C
Westlake	CALCASIEU PARISH	LA	INDORAMA VENTURES USA HOLDINGS LP (100%)	3498	C, X

(continued)

Table 1.1 (continued)

City name	County name	State	Parent companies	GHG quantity (metric tons CO_2e)	Subparts*
Alvin	BRAZORIA COUNTY	TX	INDORAMA VENTURES USA HOLDINGS LP (100%)	20	C
PORT NECHES	JEFFERSON COUNTY	TX	INDORAMA VENTURES USA HOLDINGS LP (100%)	4765	C,X
DECATUR	Morgan	AL	INDORAMA VENTURES USA HOLDINGS LP (100%)	273	C, PP
Crossett	ASHLEY COUNTY	AR	INGEVITY CORP (100%)	16	C
Covington	COVINGTON CITY	VA	INGEVITY CORP (100%)	16	C
Wever	LEE COUNTY	IA	IOWA FERTLIZER CO LLC (100%)	185	C, G, V
SODA SPRINGS	CARIBOU COUNTY	ID	ITAFOS CONDA HOLDINGS (100%)	47	C, Z
HELM	FRESNO COUNTY	CA	JR SIMPLOT CO (100%)	7	C, U, V
POCATELLO	POWER COUNTY	ID	JR SIMPLOT CO (100%)	18	C,U,Z
PASADENA	HARRIS COUNTY	TX	JX NIPPON OIL & ENERGY USA INC (100%)	21	C
ENID	GARFIELD COUNTY	OK	KOCH INDUSTRIES INC (100%)	468	C, G, PP, V
BEATRICE	GAGE COUNTY	NE	KOCH INDUSTRIES INC (100%)	107	C, G, PP, V
DODGE CITY	FORD COUNTY	KS	KOCH INDUSTRIES INC (100%)	123	C, G, PP, V
DOVER	TUSCARAWAS COUNTY	OH	KRATON CORP (100%)	10	C
KANKAKEE	KANKAKEE COUNTY	IL	Kensing LLC (100%)	17	C
Duncombe	WEBSTER COUNTY	IA	KOCH INDUSTRIES INC (100%)	108	C, G, PP, V
St. James	ST. JAMES PARISH	LA	KOCH INDUSTRIES INC (100%)	2786	C, X
Cicero	COOK COUNTY	IL	KOPPERS INC (100%)	22	C

(continued)

Table 1.1 (continued)

City name	County name	State	Parent companies	GHG quantity (metric tons CO_2e)	Subparts*
Savannah	CHATHAM COUNTY	GA	KRATON CORP (100%)	27	C
Westlake	CALCASIEU PARISH	LA	LOTTE CHEMICAL USA CORP (68.83%); WESTLAKE CHEMICAL CORP (31.17%)	63	C, X
CHEYENNE WELLS	CHEYENNE COUNTY	CO	TUMBLEWEED MIDSTREAM LLC (100%)	16	C
BAYTOWN	CHAMBERS COUNTY	TX	LANXESS CORP (100%)	38	C
EL DORADO	UNION COUNTY	AR	LANXESS CORP (100%)	25	C
EL DORADO	UNION COUNTY	AR	LANXESS CORP (100%)	14	C
EL Dorado	UNION COUNTY	AR	LANXESS CORP (100%)	10	C
PASADENA	HARRIS COUNTY	TX	LYONDELLBASELL INDUSTRIES INC (100%)	49	C, MM
CAROLINA	CAROLINA MUNICIPIO	PR	ELI LILLY & CO (100%)	32	C
EAST CHICAGO	LAKE COUNTY	IN	LINDE INC (100%)	4	C, P, PP
East Chicago	LAKE COUNTY	IN	LINDE INC (100%)	1	C, P
TEXAS CITY	GALVESTON COUNTY	TX	LINDE INC (100%)	8	C, P
WESTLAKE	CALCASIEU PARISH	LA	CONTRAN CORP (50%); VENATOR MATERIALS LLC (50%)	46	C, EE
PASADENA	HARRIS COUNTY	TX	BERKSHIRE HATHAWAY INC (100%)	11	C
PAINESVILLE	LAKE COUNTY	OH	BERKSHIRE HATHAWAY INC (100%)	16	C
DEER PARK	HARRIS COUNTY	TX	BERKSHIRE HATHAWAY INC (100%)	50	C
CHANNELVIEW	HARRIS COUNTY	TX	LYONDELLBASELL INDUSTRIES INC (100%)	209	C, MM
Hawkins	WOOD COUNTY	TX	LINDE INC (100%)	14	C
DECATUR	MORGAN COUNTY	AL	LINDE INC (100%)	3	C, P, PP

(continued)

Table 1.1 (continued)

City name	County name	State	Parent companies	GHG quantity (metric tons CO_2e)	Subparts*
Port Arthur	JEFFERSON COUNTY	TX	LINDE INC (100%)	29	C, P
GEISMAR	ASCENSION PARISH	LA	LINDE INC (100%)	1	C, P
Norco	ST. CHARLES PARISH	LA	LINDE INC (100%)	36	C, P
Westlake	CALCASIEU PARISH	LA	LYONDELLBASELL INDUSTRIES INC (50%); SASOL USA CORP (50%)	166	C, X
ST. LOUIS	ST. LOUIS CITY	MO	MALLINCKRODT ENTERPRISES HOLDINGS INC (100%)	13	C
Freeport	BRAZORIA COUNTY	TX	MEGLOBAL AMERICAS INC (100%)	40	C, PP, X
RAHWAY	UNION COUNTY	NJ	MERCK & CO INC (100%)	40	C
ELKTON	ROCKINGHAM COUNTY	VA	MERCK SHARP & DOHME LLC (100%)	20	C
CORTEZ	MONTEZUMA COUNTY	CO	MESSER NORTH AMERICA INC (100%)	1	C, PP
SAINT GABRIEL	IBERVILLE PARISH	LA	MEXICHEM FLUOR INC (100%)	15	C, L, OO
HOUSTON	HARRIS COUNTY	TX	MONUMENT CHEMICAL LLC (100%)	33	C
RITTMAN	WAYNE COUNTY	OH	STONE CANYON INDUSTRIES HOLDINGS LLC (100%)	37	C
FRIENDLY	TYLER COUNTY	WV	MOM HOLDING CO (100%)	17	C
Dickinson	DUNN COUNTY	ND	MARATHON PETROLEUM CORP (100%)	33	C
MARTINEZ	CONTRA COSTA COUNTY	CA	MARATHON PETROLEUM CORP (100%)	117	C, PP
WEST POINT	Montgomery	PA	MERCK & CO INC (100%)	118	C
Geismar	ASCENSION PARISH	LA	METHANEX USA LLC (100%)	234	C, X

(continued)

Table 1.1 (continued)

City name	County name	State	Parent companies	GHG quantity (metric tons CO_2e)	Subparts*
BRANDENBURG	MEADE COUNTY	KY	MONUMENT CHEMICAL LLC (100%); MONUMENT CHEMICAL LLC (100%)	27	C
Bartow	POLK COUNTY	FL	THE MOSAIC CO (100%)	23	C,Z
PLANT CITY	HILLSBOROUGH COUNTY	FL	THE MOSAIC CO (100%)	14	C
Lithia	HILLSBOROUGH COUNTY	FL	THE MOSAIC CO (100%)	6	C
Mulberry	POLK COUNTY	FL	THE MOSAIC CO (100%)	21	C, Z
Riverview	HILLSBOROUGH COUNTY	FL	THE MOSAIC CO (100%)	12	C, Z
Uncle Sam	ST. JAMES PARISH	LA	THE MOSAIC CO (100%)	23	C,Z
PORT ARTHUR	JEFFERSON COUNTY	TX	ARAMCO SERVICES CO (100%)	5619	C, X
Geneva	FILLMORE COUNTY	NE	NE NITRO GENEVA LLC (100%)	10	C, G
PRYOR	MAYES COUNTY	OK	Norit Americas Inc (100%)	202	C
MARSHALL	HARRISON COUNTY	TX	Norit Americas Inc (100%)	6	C
LYONS	RICE COUNTY	KS	COMPASS MINERALS INTERNATIONAL INC (100%)	25	C
MORRIS	GRUNDY COUNTY	IL	NOURYON SURFACE CHEMISTRY LLC (100%)	24	C
GEISMAR	ASCENSION PARISH	LA	NC HOLDINGS USA INC (88.46%); SABIC INNOVATIVE PLASTICS US LLC (11.54%)	223	C, X
EAST HANOVER	MORRIS COUNTY	NJ	NOVARTIS CORP (100%)	6	C
Beaumont	JEFFERSON COUNTY	TX	FIREWATER LLC (100%)	187	C, X
Rifle	RIO BLANCO COUNTY	CO	JM HUBER CORP (100%)	23	C

(continued)

Table 1.1 (continued)

City name	County name	State	Parent companies	GHG quantity (metric tons CO_2e)	Subparts*
Norco	ST. CHARLES PARISH	LA	SHELL PETROLEUM INC (100%)	5109	C, MM, X, Y
NORTH CHICAGO	LAKE COUNTY	IL	ABBVIE INC (100%)	20	C
Axis	MOBILE COUNTY	AL	NOURYON CHEMICALS LLC (100%)	11	C
COLUMBUS	LOWNDES COUNTY	MS	NOURYON CHEMICALS LLC (100%)	17	C
CONVENT	ST. JAMES PARISH	LA	OCCIDENTAL PETROLEUM CORP (100%)	39	C
GEISMAR	ASCENSION PARISH	LA	OCCIDENTAL PETROLEUM CORP (100%)	59	C, X
LUDINGTON	MASON COUNTY	MI	OCCIDENTAL PETROLEUM CORP (100%)	23	C
INGLESIDE	SAN PATRICIO COUNTY	TX	OCCIDENTAL PETROLEUM CORP (100%)	1658	C, X
Nederland	JEFFERSON COUNTY	TX	OCI PARTNERS LP (100%)	164	C, X
BAY CITY	MATAGORDA COUNTY	TX	OQ CORP (100%)	279	C
DEER PARK	HARRIS COUNTY	TX	OCCIDENTAL PETROLEUM CORP (100%)	37	C, X
LA PORTE	HARRIS COUNTY	TX	OCCIDENTAL PETROLEUM CORP (100%)	585	C, X
WICHITA	SEDGWICK COUNTY	KS	OCCIDENTAL PETROLEUM CORP (100%)	55	C
Freeport	BRAZORIA COUNTY	TX	OLIN CORP (100%)	862	C, P, X
Niagara Falls	NIAGARA COUNTY	NY	OLIN CORP (100%)	12	C
Charleston	BRADLEY COUNTY	TN	OLIN CORP (100%)	23	C
Orange	ORANGE COUNTY	TX	ORION ENGINEERED CARBONS LLC (100%)	6	C, X

(continued)

Table 1.1 (continued)

City name	County name	State	Parent companies	GHG quantity (metric tons CO_2e)	Subparts*
FRANKLIN	ST. MARY PARISH	LA	ORION ENGINEERED CARBONS LLC (100%)	2	C, X
SODA SPRINGS	CARIBOU COUNTY	ID	BAYER US HOLDING LP (100%)	185	C
PAMPA	GRAY COUNTY	TX	CABOT CORP (100%)	10	C, X
DAVENPORT	SCOTT COUNTY	IA	PB LEINER USA CORP (100%)	12	C
GEISMAR	ASCENSION PARISH	LA	PCS NITROGEN FERTILIZER LP (100%)	216	C, G, PP, V
AUGUSTA	RICHMOND COUNTY	GA	POTASH HOLDING CO INC (100%)	159	C, G, PP, V
Lima	ALLEN COUNTY	OH	POTASH HOLDING CO INC (100%)	231	C, G, PP, V
AURORA	BEAUFORT COUNTY	NC	NUTRIEN US CORP A INC (100%)	299	C, Z
KALAMAZOO	KALAMAZOO COUNTY	MI	PFIZER INC (100%)	43	C
MEMPHIS	SHELBY COUNTY	TN	PMC GROUP INC (100%)	45	C
Westlake	CALCASIEU PARISH	LA	PPG INDUSTRIES INC (100%)	42	C
Chester	DELAWARE COUNTY	PA	PQ CORP (100%)	10	C
PRYOR	MAYES COUNTY	OK	LSB INDUSTRIES INC (100%)	48	C, G, PP, V
Pasadena	HARRIS COUNTY	TX	ENTERPRISE PRODUCTS PARTNERS LP (100%)	29	C
GREENVILLE	PITT COUNTY	NC	THERMO FISHER SCIENTIFIC INC (100%)	15	C
TOLEDO	LUCAS COUNTY	OH	Perstorp Polyols (100%)	16	C
ANDOVER	ESSEX COUNTY	MA	PFIZER INC (100%)	22	C
Vega Baja	VEGA BAJA MUNICIPIO	PR	VIATRIS INC (100%)	34	C
Brunswick	GLYNN COUNTY	GA	DRT AMERICA INC (100%)	371	C
DAYTON	MONTGOMERY COUNTY	OH	TATE & LYLE INGREDIENTS AMERICAS LLC (100%)	18	C

(continued)

Table 1.1 (continued)

City name	County name	State	Parent companies	GHG quantity (metric tons CO_2e)	Subparts*
CINCINNATI	Hamilton	OH	THE PROCTER & GAMBLE CO (81.3%); DTE ENERGY CO (18.7%)	45	C
Pineville	RAPIDES PARISH	LA	THE PROCTER & GAMBLE CO (100%)	6	C
Pampa	GRAY COUNTY	TX	PROMAN USA INC (100%)	21	C, X
Seneca	LASALLE COUNTY	IL	CHEVRON CORP (100%)	13	C
DEER PARK	HARRIS COUNTY	TX	DOW INC (100%)	136	C, TT
GEISMAR	ASCENSION PARISH	LA	HUNTSMAN CORP (50%); LANXESS CORP (50%)	44	C
PEABODY	ESSEX COUNTY	MA	ROUSSELOT INC (100%)	20	C
ORANGE	ORANGE COUNTY	TX	DOW INC (100%)	11790	C, X
SWIFTWATER	MONROE COUNTY	PA	SANOFI PASTEUR INC (100%)	18	C
AIKEN	AIKEN COUNTY	SC	SAVANNAH RIVER NUCLEAR SOLUTIONS LLC (100%)	657	C
STURTEVANT	RACINE COUNTY	WI	SC JOHNSON & SON INC (100%)	34	C
GEISMAR	ASCENSION PARISH	LA	SHELL PETROLEUM INC (100%)	521	C, X
PLAQUEMINE	IBERVILLE PARISH	LA	CK TECHNOLOGIES INC (100%)	364	C, X
FREEPORT	BRAZORIA COUNTY	TX	SK CAPITAL PARTNERS LP (100%)	34	C
Orangeburg	ORANGEBURG COUNTY	SC	SI GROUP USA HOLDINGS (USHA) CORP (100%)	29	C, TT
ROCK SPRINGS	SWEETWATER COUNTY	WY	JR SIMPLOT CO (100%)	7	C, G, Z
RICEBORO	LIBERTY COUNTY	GA	SNF HOLDING CO (100%)	31	C
ANNISTON	CALHOUN COUNTY	AL	EASTMAN CHEMICAL CO (100%)	2	C

(continued)

Table 1.1 (continued)

City name	County name	State	Parent companies	GHG quantity (metric tons CO_2e)	Subparts*
ELWOOD	WILL COUNTY	IL	STEPAN CO (100%)	20	C
SAINT GABRIEL	IBERVILLE PARISH	LA	SYNGENTA CORP (100%)	29	C
HOUSTON	HARRIS COUNTY	TX	SASOL USA CORP (100%)	22	C
WESTLAKE	CALCASIEU PARISH	LA	SASOL USA CORP (100%)	354	C, X
Hugoton	TEXAS COUNTY	KS	SEABOARD ENERGY LLC (100%)	11	C
Port Allen	WEST BATON ROUGE PARISH	LA	SHELL PETROLEUM INC (100%)	21	C
Monaca	BEAVER COUNTY	PA	SHELL PETROLEUM INC (100%)	21572	C, D, DD, X
DEER PARK	HARRIS COUNTY	TX	Shell USA (100%)	6865	C, X
CHICAGO HEIGHTS	COOK COUNTY	IL	SOLVAY HOLDING INC (100%)	15	C
Baton Rouge	EAST BATON ROUGE PARISH	LA	SOLVAY HOLDING INC (100%)	18	C
TAFT	ST. CHARLES PARISH	LA	DOW INC (100%)	45540	C, X
JACKSONVILLE	DUVAL COUNTY	FL	SYMRISE INC (100%)	12	C
BRUNSWICK	GLYNN COUNTY	GA	SYMRISE INC (100%)	28	C
PACE	SANTA ROSA COUNTY	FL	EASTMAN CHEMICAL CO (100%)	18	C
SAINT GABRIEL	IBERVILLE PARISH	LA	EASTMAN CHEMICAL CO (100%)	26	C
WOODWARD	WOODWARD COUNTY	OK	CF INDUSTRIES HOLDINGS INC (100%)	184	C, G, PP, V
BISHOP	NUECES COUNTY	TX	CELANESE CORP (100%)	183	C
HOUSTON	HARRIS COUNTY	TX	SK CAPITAL PARTNERS LP (100%)	344	C
TAMPA	HILLSBOROUGH COUNTY	FL	TRADEMARK NITROGEN CORP (100%)	1	C, V
SILSBEE	HARDIN COUNTY	TX	TRECORA RESOURCES (100%)	40	C
HAMILTON	MONROE COUNTY	MS	TRONOX INC (100%)	74	C, EE

(continued)

Table 1.1 (continued)

City name	County name	State	Parent companies	GHG quantity (metric tons CO_2e)	Subparts*
St. James	ST. JAMES PARISH	LA	THE MOSAIC CO (100%)	128	C, G
McIntosh		AL	TATE & LYLE SUCRALOSE LLC (100%)	46	C
Brownsville	HAYWOOD COUNTY	TN	TEKNOR APEX (100%)	13	C
PLAQUEMINE	IBERVILLE PARISH	LA	DOW INC (100%)	11032	C, TT, X
NORTH BEND	HAMILTON COUNTY	OH	TRAMMO INC (100%)	4	C, V
PIKETON	PIKE COUNTY	OH	US GOVERNMENT (%)	5	C
SEADRIFT	CALHOUN COUNTY	TX	DOW INC (100%)	222	C, TT, X
SOUTH CHARLESTON	Kanawha	WV	DOW INC (100%)	40	C
Baton Rouge	EAST BATON ROUGE PARISH	LA	HONEYWELL UOP (100%)	12	C
SHREVEPORT	CADDO PARISH	LA	HONEYWELL INTERNATIONAL INC (100%)	26	C
Midway	GREENE COUNTY	TN	DAVIS MINING & MANUFACTURING INC (100%)	29	C, G, PP, V
CHICAGO	COOK COUNTY	IL	VANTAGE SPECIALTY CHEMICALS INC (100%)	22	C
CLAREMORE	ROGERS COUNTY	OK	CF INDUSTRIES HOLDINGS INC (100%)	450	C, G, PP, V
DARROW	ASCENSION PARISH	LA	VEOLIA ENVIRONMENTAL SERVICES NORTH AMERICA CORP (100%)	17	C
TAMAQUA	SCHUYLKILL COUNTY	PA	VERSUM MATERIALS INC (100%)	4	C, L, OO
Chicago	COOK COUNTY	IL	WR GRACE & CO (100%)	10	C
East Chicago	LAKE COUNTY	IN	WR GRACE & CO (100%)	12	C
HENNEPIN	PUTNAM COUNTY	IL	WASHINGTON MILLS GROUP INC (100%)	1	BB, C

(continued)

Table 1.1 (continued)

City name	County name	State	Parent companies	GHG quantity (metric tons CO_2e)	Subparts*
SULPHUR	CALCASIEU PARISH	LA	WESTLAKE CHEMICAL CORP (100%)	7976	C, X
SULPHUR	CALCASIEU PARISH	LA	WESTLAKE CHEMICAL CORP (100%)	76	C
GEISMAR	ASCENSION PARISH	LA	WESTLAKE CHEMICAL CORP (100%)	150	C, X
CALVERT CITY	MARSHALL COUNTY	KY	WESTLAKE CHEMICAL CORP (100%)	3282	C, X
WESTLAKE	CALCASIEU PARISH	LA	WESTLAKE CORP (100%)	30	C, X
PLAQUEMINE	IBERVILLE PARISH	LA	WESTLAKE CHEMICAL CORP (100%)	482	C, X
WESTLAKE	CALCASIEU PARISH	LA	WESTLAKE CORP (100%)	698	C, D, X
White Springs	HAMILTON COUNTY	FL	PHOSPHATE HOLDINGS INC (100%)	8	C, Z

Source Data extracted from EPA's FLIGHT tool (n.d)[3]

* Subparts: Subpart D—Electricity Generation; Subpart DD—Use of Electric Transmission and Distribution Equipment; Subpart X—Petrochemical Production; Subpart I—Electronics Manufacturing; Subpart OO—Suppliers of Industrial Greenhouse Gases; Subpart G—Ammonia Manufacturing; Subpart PP—Suppliers of Carbon Dioxide; Subpart P—Hydrogen Production; Subpart PP—Suppliers of Carbon Dioxide; Subpart U—Miscellaneous Uses of Carbonate; Subpart TT—Industrial Waste Landfills

The data available from the previous analysis shows that some of the States concentrate the highest quantity of GHG emissions in units of metric tons of carbon dioxide, which represents an important data considering that some of those states are the wealthiest in the American economy, as it is possible to check on the following Table 1.2.

As the afore table shows, the states that have a slightly higher concentration of 98% are Texas, Louisiana, Pennsylvania, Tennessee, Kentucky, Guam, Illinois, Oklahoma, Alabama, and South Carolina, as can be represented in the following geospatial analysis (Fig. 1.1).

As it is possible to appreciate in the figure, the GHG emissions are strongly concentrated in a few American States, where Texas is the main contributor with

[3] http://ghgdata.epa.gov/ghgp.

Table 1.2 Concentration index per state and total GHG quantity (metric tons CO_2e)

State	Total GHG quantity (metric tons CO_2e) per state	%	Concentration index (%)	ID10, ID15
Texas	201,884	60.50	60.50	
Louisiana	87,037	26.08	86.58	
Pennsylvania	21,807	6.54	93.12	
Tennessee	5,040	1.51	94.63	
Kentucky	3,540	1.06	95.69	
Guam	3,513	1.05	96.74	
Illinois	1,692	0.51	97.25	
Oklahoma	1,384	0.41	97.66	
Alabama	935	0.28	97.94	
South Carolina	**757**	**0.23**	**98.17**	**98.17%**
Ohio	710	0.21	98.38	
Georgia	616	0.18	98.57	
West Virginia	543	0.16	98.73	
Arizona	525	0.16	98.89	
Florida	**482**	**0.14**	**99.03**	**99.03%**
Montana	475	0.14	99.18	
Nebraska	448	0.13	99.31	
California	367	0.11	99.42	
Virginia	290	0.09	99.51	
Kansas	263	0.08	99.59	
Idaho	250	0.07	99.66	
Wyoming	154	0.05	99.71	
Michigan	130	0.04	99.75	
New Hampshire	117	0.04	99.78	
New Jersey	112	0.03	99.81	
Mississippi	97	0.03	99.84	
Indiana	80	0.02	99.87	
Delaware	73	0.02	99.89	
Maine	56	0.02	99.91	
Colorado	55	0.02	99.92	
Maryland	54	0.02	99.94	
Wisconsin	36	0.01	99.95	
Nevada	33	0.01	99.96	
Oregon	33	0.01	99.97	
Utah	32	0.01	99.98	

(continued)

Table 1.2 (continued)

State	Total GHG quantity (metric tons CO_2e) per state	%	Concentration index (%)	ID10, ID15
Washington	32	0.01	99.99	
Arkansas	13	0.00	99.99	
New Mexico	13	0.00	100.00	
New York	12	0.00	100.00	
Minnesota	1	0.00	100.00	

Source Own elaboration with data extracted from EPA's FLIGHT tool (n.d)

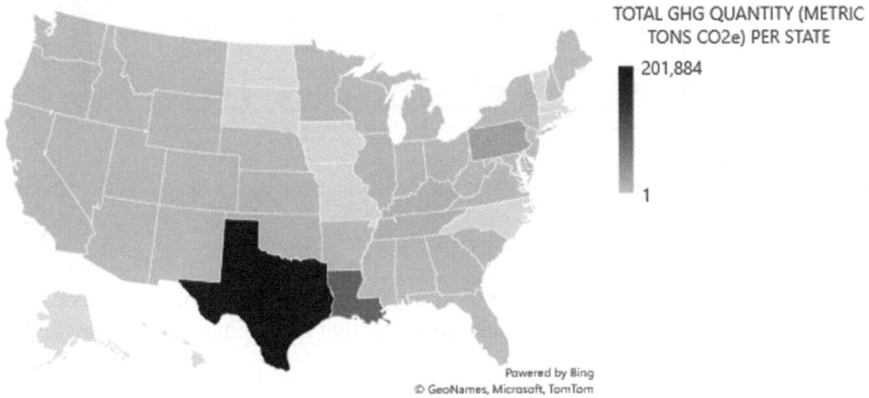

Fig. 1.1 Geospatial analysis of every American State in relation to the total GHG quantity metric tons emissions. *Source* Own ellaboration with data extracted from EPA's FLIGHT tool (n.d)

201,884; Louisiana with 87,037 and Pennsylvania with 21,807 metric tons. These American States also are considered as some of the biggest contributors in what it comes to metric tons of CO_2, as it can be seen in the following Fig. 1.2.

Those States also represent important emissions in chemicals related to GHG, with Texas (59,774), Georgia (45,540), Wyoming (14,996), and Ohio (13,547) as the regions that report the highest metric tons of CO_2e, which also represents a relevant indicator of a dense concentration of industrial activities that represents higher level of pollution to the environment. Now, in what it comes to the chemicals related to direct GHG emission of gasses reported to the Environmental Protection Agency with information from the Flight platform, where the dashboard shows the following chart,[4] for the comparison among fluorinated GHGs such as HFCs, PFCs, SF6, NF3, as well as other fully fluorinated GHGs, showing the following results (Fig. 1.3).

[4] https://www.epa.gov/.

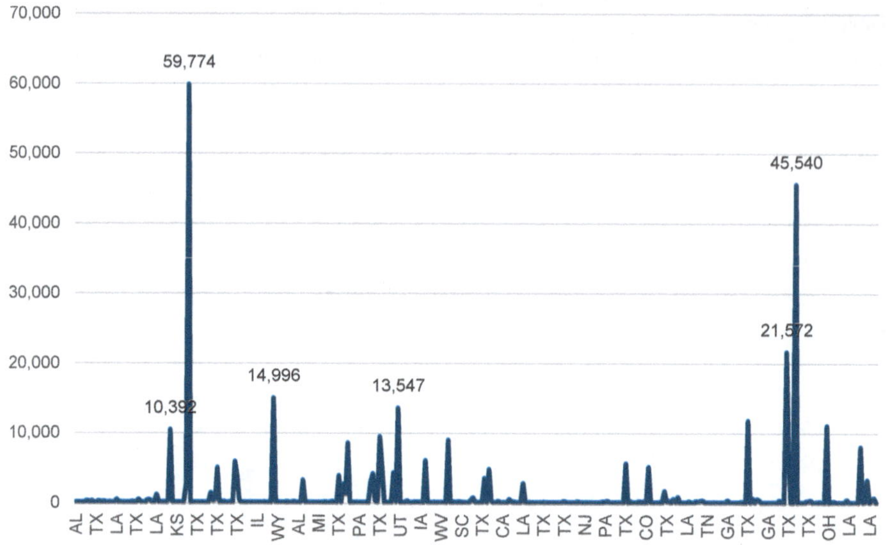

Fig. 1.2 GHG quantity (metric tons CO_2e) per American State (n.d.). *Source* Own elaboration (2024)

As it can be seen in the previous figure, the environmental impact is consistent with the regional analysis in the American States that represents an important activity in relation to the textile industry.

Discussion

As it can be seen in the previous table, the twenty companies that represent the highest contributors to the levels of Greenhouse emission gases in the textile industry, located in the USA are Bayport Polymers Llc, Dow Inc, Shell Petroleum Inc, Chevron Corp, Phillips 66, Formosa Plastics Corp USA, Dow Inc, Dow Inc, Basf Corp, Eastman Chemical Co, Ineos USA Llc, Lyondellbasell Industries Inc, Westlake Chemical Corp, Shell USA, Exxon Mobil Corp, Sabic US Holdings Lp, Chevron Corp, Phillips 66, Aramco Services Co, Shell Petroleum Inc, Lyondellbasell Industries Inc, Eastman Chemical Co, and Indorama Ventures Usa Holdings Lp; these companies together accounts for 267,565 GHG in metric tons of CO_2 emissions, which in turn represent 80.16% of the total emissions of GHG in the considered period and the considered database. It is also revealing to point out that such companies are located in 4 top tier GHG emission states, including Texas, Louisiana, Pennsylvania, and Tennessee, which represents a high concentration of such emissions in the Country, and thus, a priority in terms of focus related to protecting the cleanliness of the air.

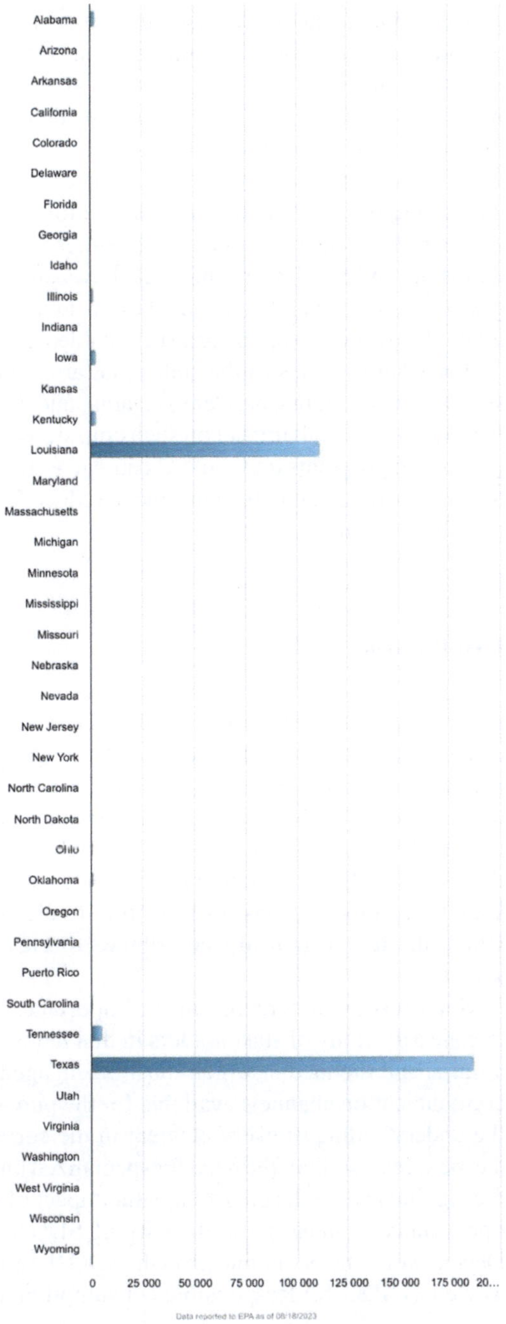

Fig. 1.3 Graphical comparison among Fluorinated GHGs such as HFCs, PFCs, SF6, NF3, as well as other fully fluorinated GHGs. *Source* Own elaboration with data obtained from EPA (n.d)[5]

[5] https://www.epa.gov/.

According to the official sources of the American government, the climate agenda established by the current administration states relevant goals such as reducing GHF emissions 50–52% below 2005 levels; also, there is an ambitious plan to achieve a net-zero emissions economy by 2050, by delivering 40% of the benefits from federal investments in climate and clean energy to disadvantaged communities[6]; thus, the analysis included in this research project shows that there are serious differences among regions in United States in relation to GHG emissions, such as in the case of Texas, Louisiana, Pennsylvania, Tennessee, Kentucky, Guam, Illinois, Oklahoma, Alabama, and South Carolina, that shows an important concentration in greenhouse gas emissions, which also should be considered as a priority to take into consideration given the overall goals toward sustainable development.

For what it comes to the states that are the top tier in terms of GHG emissions, such as Texas, Louisiana, Pennsylvania, and Tennessee, the public policies designed to reduce the overall impact on such emissions have been designed in terms of issuing permitting programs under the Clean Air Act,[7] as a proven tool to protect air quality, at the same time that will contribute to reduce GHG emissions by retaining thresholds established in the Act.

Conclusion

According to the information and the analysis presented here, given the fact that some of the American states can be considered as the major textile industry regions that register high pollution rates for GHG, this could represent a good opportunity for the development of public policies that promotes green transitions to green industries in the textile sector, establishing clear thresholds, as well as for initiatives addressing this issues at the same time that promotes competitiveness in the industry following sound recommendations for transition to clean energies for a sustainable future in where the textile industry contributes to clear solutions for the wellbeing of the society.

Such results also represent an important evidence of the evident relevance to engage a diversity of stakeholders in that industry, mostly the involvement of society at large and the media, which must be engaged in the process by maintaining clear communication channels available for the public, as well as a proper explanation for the understanding of every element in the corresponding platforms, so information can be spread without the need for specialized knowledge for the average person; also, the significance of the environmental impact of this industry represents an important opportunity to integrate a diversity of SDG´s related targets, such as the SDG 8: Decent work and economic growth; SDG 9: Industry, innovation, and infrastructure; as well as SDG 12: Responsible consumption and production.

[6] https://www.whitehouse.gov/climate/#:~:text=Reducing%20U.S.%20greenhouse%20gas%20emissions,clean%20energy%20to%20disadvantaged%20communities.

[7] https://www.epa.gov/nsr/clean-air-act-permitting-greenhouse-gases.

References

Alves L, Sá M, Cruz EF, Alves T, Alves M, Oliveira J, Santos M, Rosado da Cruz AM (2023) A traceability platform for monitoring environmental and social sustainability in the textile and clothing value chain: towards a digital passport for textiles and clothing. Sustainability 16(1):82

Arnold MG, Pfaff C, Pfaff T (2023) Circular business model strategies progressing sustainability in the German textile manufacturing industry. Sustainability 15(5):4595

Bagaria N (2022) Analysing opportunities for India in global value chains in post COVID-19 era. Foreign Trade Rev 57(3):261–282

Bagchi S, Bhattacharyya S (2021) Are export earnings from India's bilateral intra-industry trade with the US and China robust enough? Int J Dev Issues 20(3):374–392

Chowdhury NR, Paul SK, Sarker T, Shi Y (2023) Implementing smart waste management system for a sustainable circular economy in the textile industry. Int J Prod Econ 262:108876

Filho WL, Perry P, Heim H, Dinis MAP, Moda H, Ebhuoma E, Paço A (2022) An overview of the contribution of the textiles sector to climate change. Front Environ Sci 10:973102. https://doi.org/10.3389/fenvs.2022.973102

Gallico D (2020) The cotton value chain: improving its sustainability, inclusiveness and value adding capabilities: Italy-Egypt collaborative sustainability-based innovation. Sustainability awareness and green information technologies. Springer International Publishing, Cham, pp 199–214

Gnidchenko AA (2023) The links of Russia with EU, USA, China and ASEAN in international supply chains. HSE Econ J 27(4):527–549 (In Russ.)

Gomes K, Caucci S, Morris J, Guenther E, Miggelbrink J (2024) Sustainability transformation in the textile industry—the case of wastewater management. Bus Strat & Dev 7(1):e324

Imran S, Mujtaba MA, Zafar MM, Hussain A, Mehmood A, Farwa UE, Korakianitis T, Kalam MA, Fayaz H, Saleel CA (2023) Assessing the potential of GHG emissions for the textile sector: a baseline study. Heliyon 9(11). https://doi.org/10.1016/j.heliyon.2023.e22404

Joa B, Hottenroth H, Jungmichel N, Schmidt M (2014) Introduction of a feasible performance indicator for corporate water accounting–a case study on the cotton textile chain. J Clean Prod 82:143–153

Joséphine R, Andrée-Anne L, Samir L (2022) Lifecycle traceability towards sustainable and circular value chains: analysis framework and state of the art in the fashion industry. IFAC-PapersOnLine 55(10):1705–1710

Ju X, Ma Z, Chen F, Yao P, Liu J, Sriboonchitta S (2023) Environmental regulation, industrial agglomeration, and sustainable development in the Chinese textile industry. Technol Econ Dev Econ 29(6):1633–1656

Köhler AR, Som C, Helland A, Gottschalk F (2008) Studying the potential release of carbon nanotubes throughout the application life cycle. J Clean Prod 16(8–9):927–937

Løvbak Berg L, Klepp IG, Sigaard AS, Broda J, Rom M, Kobiela-Mendrek K (2023) Reducing plastic in consumer goods: opportunities for coarser wool. Fibers 11(2):15

Luján-Ornelas C, Güereca LP, Franco-García ML, Heldeweg M (2020) A life cycle thinking approach to analyse sustainability in the textile industry: a literature review. Sustainability 12(23):10193

Manshoven S, Van Opstal W (2022) The carrot or the stick? Stakeholder support for mandatory regulations towards a circular fashion system. Sustainability 14(22):14671. https://doi.org/10.3390/su142214671

Olofsson L, Mark-Herbert C (2020) Creating shared values by integrating UN sustainable development goals in corporate communication—the case of apparel retail. Sustainability 12(21):8806

Palacios-Mateo C, Van der Meer Y, Seide G (2021) Analysis of the polyester clothing value chain to identify key intervention points for sustainability. Environ Sci Eur 33(1):2

Pande B, Adil GK (2022) Assessment of the current state of sustainability in a manufacturing firm. Int J Product Perform Manag 71(4):1254–1276

Pellegrini M, Guzzini A, Saccani C (2023) Sustainability of medical gowns supply chain management in the post-covid era. In Proceedings of the changeable, agile, reconfigurable and virtual production conference and the world mass customization & personalization conference, pp 469–476. Springer International Publishing, Cham

Peters G, Li M, Lenzen M (2021) The need to decelerate fast fashion in a hot climate-a global sustainability perspective on the garment industry. J Clean Prod 295:126390

Singh J, Sung K, Cooper T, West K, Mont O (2019) Challenges and opportunities for scaling up upcycling businesses–the case of textile and wood upcycling businesses in the UK. Resour Conserv Recycl 150:104439

Stellmach D, Weiß M, Seibold J, Tilebein M (2022) Towards a digital workflow solution for cradle-to-gate sustainability information in textile value chains. In Herberger D, Hübner M (eds) Proceedings of the conference on production systems and logistics: CPSL 2022, pp 723–732. Hannover Publishing. https://doi.org/10.15488/12155

Thanki S, Thakkar J (2018) A quantitative framework for lean and green assessment of supply chain performance. Int J Product Perform Manag 67(2):366–400

TTRI Taiwan Textile Research Institute (2013) Low carbon intelligent operations for textile industry in APEC economies. https://www.apec.org/docs/default-source/Publications/2013/8/Low-Carbon-Intelligent-Operations-for-Textile-Industry-in-APEC-Economies---Project/TOC/Main-Report.pdf

United Nations Environment Programme (2023) Sustainability and circularity in the textile value chain–a global roadmap. Paris. https://circulareconomy.europa.eu/platform/sites/default/files/2023-12/Full%20Report%20-%20UNEP%20Sustainability%20and%20Circularity%20in%20the%20Textile%20Value%20Chain%20A%20Global%20Roadmap_0.pdf

Zeying Y (2014) Analysis of fiscal and financial support models for strategic emerging enterprises in different innovation stages. Manag Mod 34(2):54–56

Chapter 2
Sustainable Practices in the Textile Sector: Case Study of the Global Value Chain of the American Economies

Aníbal Enrique Toscano-Hernández and Saúl Alfonso Esparza-Rodríguez

Introduction

In general terms, production processes organization requires identifying essential parts of a set of linked activities that contribute to the shifting nature of firms and nations in a global production network, generally represented by cost-competitive advantages, availability of new tradable commodities and other factors that contribute to enhance competitive advantages (Ahmad and Primi 2017).

In such a scenario, the textile value chains represent a complex structure of linked collaboration between firms and economies, where supporting forces from a diversity of stakeholders are relevant to design, create, maintain, and improve global interlinked activities where every part of the chain represent real opportunities to share and scale coordination mechanisms for cohesion and supporting efforts to minimize negative impacts on nature, people and economies.[1] For what it comes to governance mechanisms for addressing sustainable practices, the inclusion of public governance, encompassing collective bargaining agreements, and multi-actor global binding agreements, will contribute to effectively addressing essential working deficits (Anner 2021) It is relevant to highlight that global textile and apparel value chain is forecast to grow at a CAGR of 5% per year and reach US$2,100bn by 2025 (Lu 2015).

[1] https://www.unep.org/events/unea/connecting-threads-coordinated-policy-response-transform-textile-value-chain-and-offer#:~:text=The%20textile%20value%20chain%20is%20complex%20and,and%20provide%20opportunities%20to%20share%20and%20scale.

A. E. Toscano-Hernández (✉)
Universidad del Sinú, Montería, Colombia
e-mail: anibaltoscano@unisinu.edu.co

S. A. Esparza-Rodríguez
National Autonomous University of Mexico, Mexico City, Mexico

A. E. Toscano-Hernández et al. (eds.), *Sustainability, Stakeholders and Marketing in the Textile Sector*, SDGs and Textiles, https://doi.org/10.1007/978-981-97-6952-0_2

Given its relevance as an essential factor for sustainable textile value chains, the role of different stakeholders stands out as a significant issue in supporting ecologically responsible, economically profitable, and socially responsible GVCs, where the consumer is making decisions using emotion-driven and aspirational messages to engage with companies. Brands and retailers engage and inform consumers to shift them to sustainable alternatives. Policymakers creating policies and incentives that help to shift consumption patterns and disincentivize waste. NGOs and technical organizations engage consumers in shifting behavior to a sustainable pathway to reducing consumption, contributing to change patterns across different regions and demographics, consumer attitudes, and behavior change (United Nations Environment Programme 2023); In those relationships, complexities arise where one stakeholder is in a strong position for changing economic and political factors by altering production and consumption patterns (Jones et al. 2019).

Literature Review

The participation in the global textile and apparel value of interrelated economies has effect in high domestic share on exports, also showing effects in tariffs, Foreign Direct Investment, and labor costs as well, where different tasks are performed considering whether the country has a developing economy (primarily manufacturing) or developed economies (branding, design) (Rodil-Marzábal et al. 2022); Up to this day, global value chains (GVCs), led by transnational corporations (TNCs), have reshaped the textile and apparel industries that generate wild competition at the lower value-added stages of production, (Aguiar de Medeiros and Trebat 2017).

Generally speaking, there are some schools of thought that contribute to the study of industrial organization patterns and strategic decisions of participants in the textile industry value chain that describe the textile industry as a network-based approach capable of ranging multiple spatial scales while elucidating the dynamic relationships emphasizing the increasing importance of location and institutional actors to increase competitiveness, demanding essential raw materials for production in the textile industry, exploitation cycles that contributes to environmental and socio-labor degradation, hierarchies and inequalities. (Márquez and Vásquez 2020); Also, some conditions contribute to the level of competitiveness of a given GVC, such as neighboring economies, market niches, upgrading indicators, export markets, foreign supplies and trade agents, (Evgeniev and Gereffi 2008), all of them elements confirming the specific environment of textile industry firms.

The COVID-19 pandemic created uncertainties that force clothing and textile (C&T) companies to assess value chain segments that have been affected due to the crisis, and to implement appropriate strategies during times of crisis (Zhao and Kim 2021); Such period of fragmentation and internationalization of production, has been considered in several previous studies by bridging a 'micro–macro gap' in development theory, meaning making valid inferences from micro-level actors (firms) to macro-sociological outcomes; and second, reconciling its firm-level organizational approach with institutionalism (Dallas 2014).

As a potential answer to such issues, sustainability practices have arisen as a point of reference focused on examining whether and how shape relevant factors in the textile GVC such as labor conditions, factory conditions, and government labor inspection (Bartley and Egels-Zandén 2015). In that sense, maintaining distinct competitive advantages relies on the ability of MNC and GVC to agile respond to consumer tastes and trends, seeking economies of scale cultivate supplier networks that contain prices, and keeping pace with marketing platforms (Anwar 2017).

In the recent era, sustainability reporting regards a vital tool for identifying social and environmental impacts in an organization, also representing a critical factor that helps reducing risk and increase small businesses' competitiveness, facilitating integration in global value chains, facilitating companies to improve their vision, strengthen linking activities with dominant companies, and generating greater competitiveness by accessing new markets (Bunclark and Barcellos-Paula 2021). To be able to have access to such opportunities, the macroenvironment must provide resources, state planning, coordination, resilience to external shocks, and sustained performance of firms as well (Ngo 2017). As an added benefit, economies can improve their competitive position from the advantages of GVC participation by upgrading a position that allows them to reap benefits of spillovers to a given domestic economy (Ozer et al. 2016). Also, some other competitive advantages for economies joining textile GVCs come from textile material supply, full-package production, lean manufacturing, efficient transportation, logistics services, lower labor costs, production differentiation, and specialization as well (Lu and Karpova 2011). In such complex environment, some drivers that are mobilized by stakeholders to adapt to a textile and apparel value chain structure are based in the creation of scarce assets of knowledge throughout the entire value chain to attain the sector strategic reconfiguration, where the roles that must be undertaken by government, businesses, institutions, and society are essential to support the sector profile transitions to sustainable practices as a competitive advantage (Bruno et al. 2009).

For what it comes to the Americas, Mexico and Central America have been end market diversification in terms of upgrading trajectories of leading apparel exporters adapting to the end of textile and apparel quotas; In the case of Mexico, regional trade agreements have provided preferential access to the US market and ties to brand manufacturers, creating a reliance on US exports in the process (Frederick and Gereffi 2011; Frederick and Cassill, 2009) The effects of trade liberalization on employment and the labor market in Mexico's manufacturing industry shows relevant links between international trade and some indicators of sustainable practices such as decent work (Moreno-Brid et al. 2023). In the other hand, for Brazil the clothing sector presents some working conditions promoted by new dynamics in the forms of precariousness that have always marked the sector (Leite et al. 2017).

The Sustainable Value Chain Model as a Practice that Promotes Competitiveness

In recent years, sustainability in the value chain has received increased attention, from both practitioners and scholars, mainly due to the significant importance given by consumers, profit and non-for-profit organizations, as well as other relevant stakeholders such as local communities, and government as well, striving for achieving the triple bottom line model (including social, environmental and economic benefits) (Govindan et al. 2013).

Following that theoretical framework, some recent global events such as the COVID-19 pandemic evidenced the relevance of including short-term environmental sustainability gains as relevant goals for the strategy of the organization, where sustainability and resilience are complementary objectives, that could be achieved under certain conditions such as discretionary decisions from the owners, providing incentives, including knowledge, and needed resources to invest in the sustainable future of the firm (Le Breton–Miller and Miller 2006), creating a pathway where environmental sustainability practices can help organizations manage in this and future competitive contexts, embracing some social implications such as broad economic, operational, social, and ecological-environmental sustainability (Sarkis 2020).

Nowadays, the market is following trends for sustainable supply chain management, based on sustainable practices in purchasing and supply to reduce negative impacts on the environment, economy and society, where companies acknowledged as leaders in their respective industries can be classified in a typology tha considers internal focusers (for internal and organic growth), reserved players (reactive to the change of the environment), external responders (externally focused), and agenda setters (generally leaders in their fields) (Walker and Jones 2012); such categories also represent a useful resource to identify how companies are engaging with SDG's targets that are actionable within their value chain operations, looking to avoid harm, trying to do good (Van Zanten and Van Tulder 2018) by taking into consideration some contingency variables such as power, dependency distance, industry, and knowledge resources, creating a multi-tier sustainable supply chain management (Tachizawa and Yew Wong 2014).

In such environment, the potential competitive advantage for firms relies on the tactics used in the creation and management of a sustainable supply chain, increasing their competitive advantage by integrating the social, economic, and environmental framework as an essential part of the strategy, from a natural-resource-based view of the firm perspective, is supported by accountability, strategy and green logistics (Markley and Davis 2007). Such a structural model is relevant for performance measures and drivers, in terms of intra- and inter-organizational environmental practices, performance measures, and institutional drivers to promote sustainable industrial development (Guang Shi et al. 2012).

Since most sustainable supply chains are situated in dynamic environments, management requires the application of a dynamic management approach, where sustainable supply chain management practices require agility of the overall supply chain to lead to a higher performance (Beske 2012). Given the relevance of those practices for competitiveness, some countries are promoting good practices in improving the social sustainability of their own supply chains (Garments 2014), creating a sustainable economic agenda for regional competitiveness.

Considering the aforementioned information, sustainable practices in value chain designed to promote competitiveness must be considered as a complex multidimensional construct, build around a set of variables for regions to be considered sustainable in promoting good environmental practices, and thus, the next section explains the chosen methodology to explain the research problem with interactions of a set of latent variables.

Methodology

The present research uses a structural equation modeling with the support of the SmartPLS[2] software, using data for the global value chain of the textile industry provided by the United Nations, Department Of Economic and Social Affairs Statistics, Division Trade Statistics, available at the ComTrade database for such institution.[3] This database covers a global trade data platform that includes statistics for 200 nations, more than 99% of merchandise trade, imports and exports of goods and services, with data filtered with the period corresponding to 2023, including trade flow related to import (M) and export (X) included in the Commodity code considered in Chapter 58 of the International Trade Classification (ITC). Special woven fabrics, tufted textile fabrics, lace, tapestries, trimmings, and embroidery (not apply to textile fabrics that are impregnated, coated, covered, or laminated, or to other goods in Chapter 59)[4] (United States International Trade Commission 2011), considering the total model of transport (air, water, sea, inland waterway, land, railway, road, among others), in terms of total trade value, considering the World as 2nd partner in general terms, as well as among the same economies in the Americas as an specific analysis.

The present analysis covers the 10 most important economies of the Americas: United States of America, Canada, Mexico, Brazil, Argentina, Colombia, Peru, Chile, Ecuador, and Puerto Rico, according to the GDP reported to the World Bank platform. The presentation of information includes descriptive statistics, descriptive tables, figures, and a statistically proven model based on structural equation modeling with partial least squares.

---

[2] https://www.smartpls.com/.

[3] https://comtradeplus.un.org/.

[4] https://www.wcoomd.org/~/media/wco/public/global/pdf/topics/nomenclature/instruments-and-tools/hs-nomenclature-older-edition/2007/hs-2007/1158_2007e.pdf?la=en.

Table 2.1 Strongest economies in the Americas according to the GDP (2022)[5]

Country	Most recent value (millions)
United States of America	$25,439,700.00
Canada	$2,161,483.37
Brazil	$1,920,095.78
Mexico	$1,465,854.09
Argentina	$631,133.38
Colombia	$343,622.11
Chile	$301,024.72
Peru	$242,631.57
Ecuador	$115,049.48
Puerto Rico	$113,434.80

Source Own elaboration with information from the World Bank (n.d.)

Results

First, to identify the countries that will be included in the study we are performing a descriptive analysis regarding the most recent value in millions of American dollars for the Gross Domestic Product of the strongest economies in the Americas; this information is essential to perform an analysis based on those economies that represents the highest value in economic transaction in the area, as follows (Table 2.1).

As the previous table shows, the listed economies in order of transaction size are the United States of America, Canada, Brazil, Mexico, Argentina, Colombia, Chile, Peru, Ecuador, and Puerto Rico; the American economies in the continent are noticeable higher, representing a comparative 78% of value considering such economies, followed by Canada, Brazil and Mexico, with approximately 7, 6 and 4% each, three of those economies are part of the USMCA agreement.

The next table shows the total trade of the imports of American economies with economies in the World, corresponding to the Commodity code considered in Chapter 58 of the International Trade Classification (ITC): Special woven fabrics, tufted textile fabrics, lace, tapestries, trimmings, and embroidery (not apply to textile fabrics that are impregnated, coated, covered, or laminated, or to other goods in Chapter 59), including total modes of transportation, with Economies of the World as second partner (Expressed in concentration ratio); thus, the analysis is being performed on the value generated in the products related to Chapter 58, for a proper analysis with a clear scope, as it is showed next (Table 2.2).

The former table shows an important level of concentration of the total trade reported in the United Nations Comtrade database, with a strong participation in the first 10 cases, including Canada, Mexico, Brazil, Nicaragua, El Salvador, Paraguay,

[5] https://data.worldbank.org/indicator/NY.GDP.MKTP.CD?locations=US&most_recent_value_desc=true.

Table 2.2 Total trade of the imports of American economies corresponding to the Commodity code considered in Chapter 58 of the International Trade Classification (ITC) including total modes of transportation, with Economies of the World as second partner (Expressed in concentration ratio)

Reporter	Partner	Trade Value (US$)	%	Concentration ratio	CK index
Mexico	Canada	$68,004,776	59.1		
Mexico	Brazil	$11,374,234	9.9		
Nicaragua	El Salvador	$4,588,229	4.0		
Paraguay	Brazil	$4,128,193	3.6		
Nicaragua	Honduras	$3,919,988	3.4		
Canada	Mexico	$2,420,456	2.1		
Guatemala	Honduras	$2,259,266	2.0		
Mexico	El Salvador	$2,060,331	1.8		
Guatemala	El Salvador	$2,050,920	1.8		
Brazil	Paraguay	$1,235,936	1.1	88.7%	CK10

Source Own elaboration with information from the UN Comtrade (n.d.)

and Guatemala. This data excludes the participation of USA since this economy dominates economically the region, and hence, this will provide a clear scenario for other relevant economies in terms of economic trade for imports.

In the cases of the total trade of the exports of American economies with economies in the World, corresponding to the Commodity code considered in Chapter 58 of the International Trade Classification (ITC): Special woven fabrics, tufted textile fabrics, lace, tapestries, trimmings, and embroidery (not apply to textile fabrics that are impregnated, coated, covered, or laminated, or to other goods in Chapter 59), including total modes of transportation, with Economies of the World as second partner (Table 2.3).

For the data shown in the previous table, the level of concentration is wider, including a concentration of the 10 most relevant cases with 82.3%, but 90.2% concentration rate for the 18 first cases; in such scenario, the commerce includes some other countries beyond the Americas (such as in the case of Netherlands and Portugal), as well as smaller economies such as Bolivia and Dominican Republic (Table 2.4).

This concentration shown in the former table also is present in the export trade flow among American Economies in the textile industry, in terms of Trade Value (US$), as it can be seen in the next Table 2.5.

Once the relevance of these economies in terms of commercial trade, imports, and exports among American economies related to Chapter 58 is been determined, the next figures will show some relevant variables related to sustainability in the area, being the first one the share of public cash transfers received by working-age individuals in low and high-income groups, including American economies and the comparative indicator with the OECD average, as follows (Fig. 2.1).

As it is shown in the precedent figure, the economies considered in the share of public cash transfers received by working-age individuals in low and high-income

Table 2.3 Total trade of the exports of American economies with economies in the World, corresponding to the Commodity code considered in Chapter 58 of the International Trade Classification (ITC) including total modes of transportation, with Economies of the World as second partner

Reporter	Partner	Trade Value (US$)	%	Concentration ratio	CK index
Canada	Mexico	$49,695,229	40.6		
Brazil	Mexico	$10,084,347	8.2		
Germany	Mexico	$9,255,321	7.6		
Brazil	Argentina	$7,032,487	5.8		
Guatemala	El Salvador	$4,822,834	3.9		
Germany	Brazil	$4,720,565	3.9		
Guatemala	Honduras	$3,970,447	3.2		
Brazil	Paraguay	$3,776,924	3.1		
Netherlands	Canada	$3,734,699	3.1		
Mexico	Nicaragua	$3,549,579	2.9	82.3%	CK10
Germany	Canada	$1,753,340	1.4		
Guatemala	Nicaragua	$1,564,972	1.3		
Mexico	Canada	$1,468,260	1.2		
Paraguay	Brazil	$1,226,372	1.0		
Brazil	Bolivia	$1,023,550	0.8		
Mexico	El Salvador	$982,488	0.8		
Portugal	Dominican Rep	$950,325	0.8		
Mexico	Guatemala	$721,865	0.6	90.2%	CK18

Source Own elaboration with information from the UN Comtrade (n.d.)

Table 2.4 Import trade flow among American Economies in the textile industry, in terms of Trade Value (US$)

Partner	Trade Value (US$)	%	Concentration ratio	CK
USA	$99,417,306	92.8		
Brazil	$4,125,817	3.9		
Mexico	$966,157	0.9		
Mexico	$966,157	0.9		
Mexico	$313,266	0.3		
Canada	$205,044	0.2		
Canada	$205,044	0.2		
Canada	$202,186	0.2		
Mexico	$187,023	0.2		
Brazil	$172,392	0.2		
Brazil	$169,403	0.2	99.20%	CK11

Source Own elaboration with information from the UN Comtrade (n.d.)

Table 2.5 Export trade flow among American Economies in the textile industry, in terms of Trade Value (US$)

Partner	2nd partner	Trade Value (US$)
Argentina	American Economies	$476,413
Brazil	American Economies	$927,157
Canada	American Economies	$5,835,716
Chile	American Economies	$475,014
Colombia	American Economies	$666,389
Ecuador	American Economies	$109,482
Mexico	American Economies	$2,868,973
Peru	American Economies	$147,306
USA	American Economies	$984,266.00

Source Own elaboration with data retrieved from the UN Comtrade Dataset (n.d.)[6]

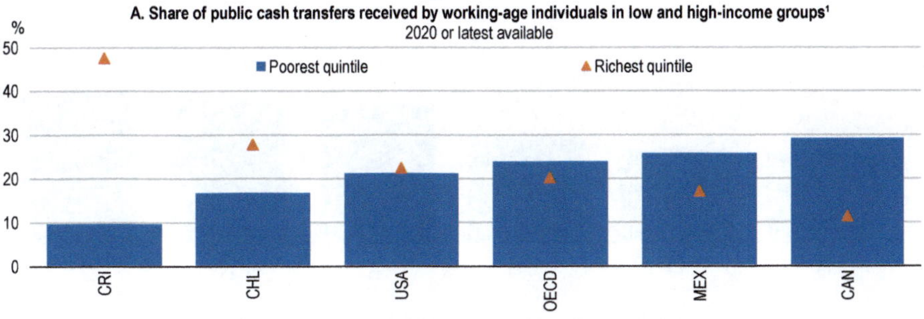

Fig. 2.1 Share of public cash transfers received by working-age individuals in low and high-income groups. *Source* Own elaboration with data from the OECD library (n.d.)[7]

groups include Costa Rica, Chile, USA with a noticeable difference in cash transfer per type of group, meanwhile Mexico and Canada, have a shorter discrepancy between those two groups.

For what it comes to gender inequalities by employment gap in American Economies, the following figure includes the cases for Canada, USA, Chile, Argentina, Brazil, Costa Rica, Colombia and Mexico, from lower to higher, as it can be seen next (Fig. 2.2).

The precedent figure shows a relevant difference between Canada (5% approximately), and Mexico (roughly above 30%) in terms of inequalities; also, only Canada and USA have lower percentages compared with the OECD average.

---

[6] https://comtradeplus.un.org/TradeFlow?Frequency=A&Flows=X&CommodityCodes=58&Partners=0&Reporters=842&period=2023&AggregateBy=none&BreakdownMode=plus.

[7] https://www.oecd-ilibrary.org/sites/e02dd0a9-en/index.html?itemId=/content/component/e02dd0a9-en.

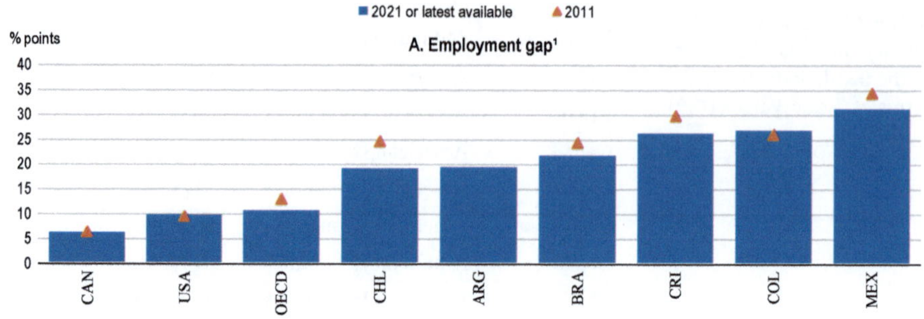

Fig. 2.2 Gender inequalities by employment gap in American Economies. *Source* Own elaboration with data from the OECD library (n.d.)[8]

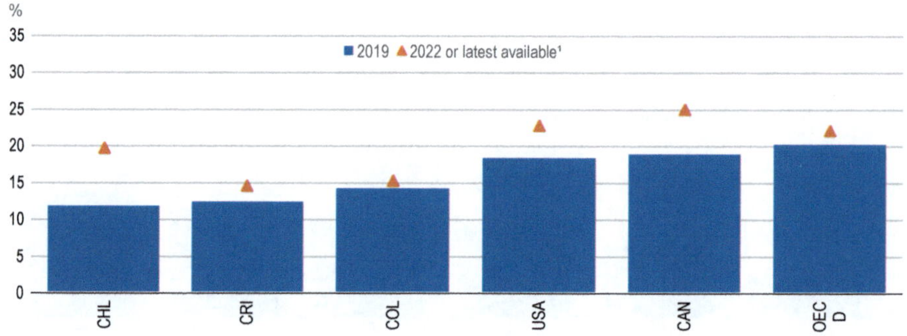

Fig. 2.3 Public social expenditure as a percentage of GDP in American Economies. *Source* Own elaboration with data from the OECD library (n.d.)[9]

The next figure shows the public social expenditures as a percentage of GDP in American Economies, including the cases for Chile, Costa Rica, Colombia, USA, and Canada compared with the OECD average, as the next figure shows (Fig. 2.3).

The comparative graphic shows that all the included economies are lower in terms of Public social expenditure as a percentage of GDP in American Economies compared to the OECD average, with the United States and Canada nearer to such an average. For what it comes to the green bonds issuance as a percentage of GDP, the statistics shows the following (Fig. 2.4).

As it can be seen in the previous figure, the only economy that has a higher percentage compared to the OECD average is Chile, with over 14% of proportional share compared to GDP. The next figure relates to carbon pricing score for the most

[8] https://www.oecd-ilibrary.org/sites/e02dd0a9-en/index.html?itemId=/content/component/e02dd0a9-en.

[9] https://www.oecd-ilibrary.org/sites/e02dd0a9-en/index.html?itemId=/content/component/e02dd0a9-en.

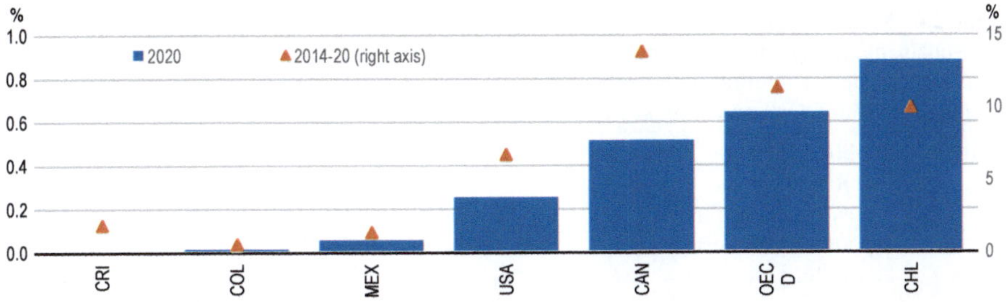

Fig. 2.4 Green bonds issuance as a percentage of GDP. *Source* Own elaboration with data from the OECD library (n.d.)[10]

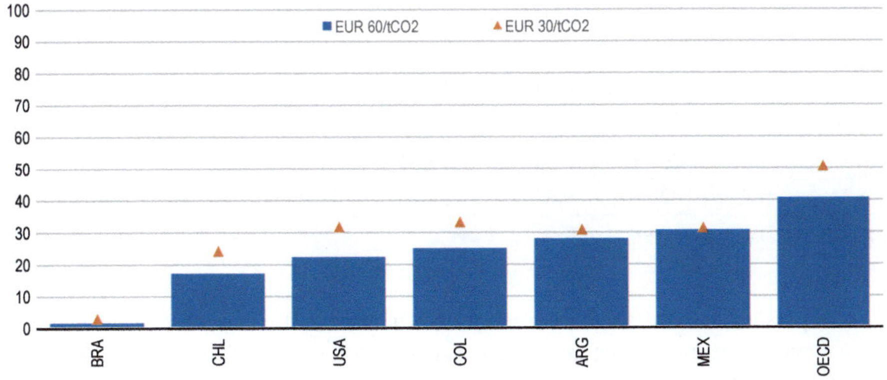

Fig. 2.5 Carbon pricing score,[1] 2018. *Source* Own elaboration with data from the OECD library (n.d.)[11]

recent data, in comparison to the EURO as the comparative currency included in the database of the OECD, as it can be seen next (Fig. 2.5).

The precedent figure shows that all the American economics included in the graphic are below OECD with only Mexico being near to such indicator. Using this comprehensive data, the next part of the analysis regards a multivariable regional analysis for sustainable economies of the Americas in the global value chain, including Public cash transfers (Poorest Quintile), Public cash transfers (Richest quintile), Employment gap (%), Public social expenditure, Green bonds issuance (%GDP), Carbon Emission Pricing Score (30/tCO_2), and Context Sustainability, so it is possible to start a contrast analysis among the American economies in the selected textile sector, as follows (Table 2.6).

[10] https://www.oecd-ilibrary.org/sites/e02dd0a9-en/index.html?itemId=/content/component/e02dd0a9-en.

[11] https://www.oecd-ilibrary.org/sites/e02dd0a9-en/index.html?itemId=/content/component/e02dd0a9-en.

Table 2.6 Multivariable regional analysis for sustainable economies of the Americas in the global value chain, including public cash transfers (poorest quintile), public cash transfers (richest quintile), employment gap (%), public social expenditure, green bonds issuance (%GDP), carbon emission pricing score (30/tCO_2), and context sustainability

Partner	Public cash transfers (poorest quintile)	Public cash transfers (richest quintile)	Employment gap (%)	Public social expenditure	Green bonds issuance (%GDP)	Carbon emission pricing score (30/tCO_2)	Context sustainability
Argentina	23.85	20.13	19.29	22.04	0.64	30.55	0.44
Brazil	23.85	20.13	21.56	22.04	0.64	2.63	0.10
Canada	29.02	11.29	6.15	24.91	0.51	50.38	−0.71
Chile	16.76	27.68	19.02	19.61	0.88	23.80	0.61
Colombia	23.85	20.13	26.76	15.15	0.01	32.87	−0.10
Ecuador	23.85	20.13	10.54	22.04	0.64	50.38	0.00
Mexico	25.70	16.98	31.09	22.04	0.05	31.03	0.57
Peru	23.85	20.13	10.54	22.04	0.64	50.38	0.00
USA	21.15	22.33	9.64	22.70	0.25	31.36	−1.05

Source Own elaboration (2024)

To represent the contrast among economies and the multivariable regional analysis for sustainable economies of the Americas in the global value chain, including Public cash transfers (Poorest Quintile), Public cash transfers (Richest quintile), Employment gap (%), Public social expenditure, Green bonds issuance (%GDP), Carbon Emission Pricing Score (30/tCO_2), and Context Sustainability, the following graphic includes every economy and variables, with the green bonds issuance and context sustainability as contrasting lines (Fig. 2.6).

The descriptive analysis can be used to establish the possibility of a non-parametric correlation among the multivariable regional analysis for sustainable economies of

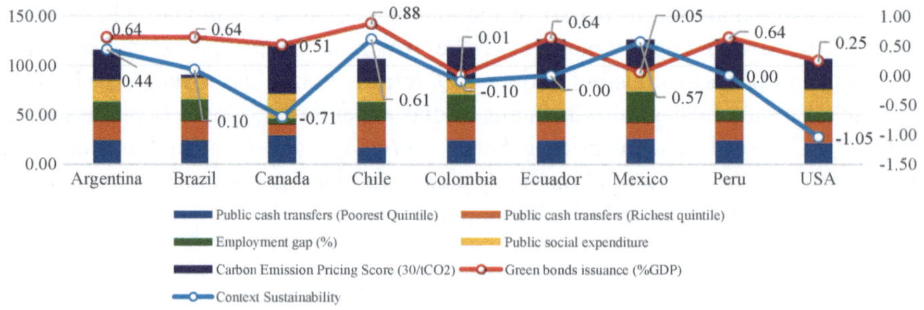

Fig. 2.6 Multivariable regional analysis for sustainable economies of the Americas in the global value chain. *Source* Own elaboration (2024)

the Americas in the global value chain, so the potential relationship between variables can be interpreted as pairing variables, as follows (Table 2.7).

The non-parametric correlation analysis using the Spearman test shows that the only statistically valid correlation is in terms of the Trade Value, the public cash transfers (Poorest Quintile), and Public cash transfers (Richest quintile); while the remaining variables have no significant correlation, the trade value of the textile industry in the American economies show a positive relation with the public cash

Table 2.7 Correlation analysis

	Trade Value (US$)	Public cash transfers (poorest quintile)	Public cash transfers (richest quintile)	Employment gap (%)	Public social expenditure	Green bonds issuance (%GDP)	Carbon emission pricing score (30/ tCO_2)
Trade Value (US$)	1.00	**0.679***	**−0.805****	−0.17	0.49	−0.24	0.24
Significance		0.04	0.01	0.67	0.18	0.54	0.53
Public cash transfers (poorest quintile)	**0.679***	1.00	**−0.981****	−0.11	0.40	−0.36	0.39
Significance	0.04		0.00	0.78	0.29	0.34	0.30
Public cash transfers (richest quintile)	**−0.805****	**−0.981****	1.00	0.17	−0.47	0.35	−0.41
Significance	0.01	0.00		0.66	0.20	0.36	0.27
Employment gap (%)	−0.17	−0.11	0.17	1.00	−0.58	−0.46	−0.56
Significance	0.67	0.78	0.66		0.10	0.21	0.11
Public social expenditure	0.49	0.40	0.47	0.58	1.00	0.33	0.23
Significance	0.18	0.29	0.20	0.10		0.38	0.56
Green bonds issuance (%GDP)	−0.24	−0.36	0.35	−0.46	0.33	1.00	−0.04
Significance	0.54	0.34	0.36	0.21	0.38		0.92
Carbon emission pricing score (30/ tCO_2)	0.24	0.39	−0.41	−0.56	0.23	−0.04	1.00
Significance	0.53	0.30	0.27	0.11	0.56	0.92	

* Correlation is significant at the 0.05 level (2-tailed)
** Correlation is significant at the 0.01 level (2-tailed)
Source Own elaboration (2024)

transfers for the poorest quintile, and conversely a negative correlation with the public cash transfers for the richest quintile, giving the idea that the sector could represent a contributor for public policies being transferred to the poorest population, while it has a negative correlation with the richest quintile, which can represent less subsidies and benefits for that sector of the population.

Structural Equation Modeling (Partial Least Squares)

Structural equation modeling (SEM) is a family of statistical techniques that has become very popular in business and social sciences. Its ability to model latent variables, to take into account various forms of measurement error, and to test entire theories makes it useful for a plethora of research questions. Two types of SEM can be distinguished: covariance- and variance-based SEM. Covariance-based SEM estimates model parameters using the empirical variance–covariance matrix, and it is the method of choice if the hypothesized model consists of one or more common factors (Henseler et al. 2016). Thus, for this project, the PLS-SEM model was performed, using the SmartPLS software 4 version, which is the updated version to perform the needed analysis.

According to the recommended analysis process in Henseler et al. (2016), if the program shows a consistent construct correlation matrix, then it is possible to perform the path coefficient estimates, showing the following results (Table 2.8).

As it is possible to observe, the direct interactions among latent variables show a positive connection between the carbon emission pricing score and context sustainability (which is a construct calculated by comparing the overall results of every economy against the weighted average of the remaining variables); as well as the employment gap and the context sustainability, as well as green bond issuance an context sustainability; conversely the public social expenditure has a negative path with the employment gap, which is an indicator of a lack of investment in reducing such gap in the American economies specifically in the textile industry (Table 2.9).

As it is possible to observe in the previous table, the interactions that are statistically significant among latent variables such as the public social expenditure and the context sustainability are mediated by the employment gap, which is an indicator

Table 2.8 Path coefficients

Interactions between latent variables	Path coefficients
CEP -> CS	0.494
EG -> CS	1.358
GBI -> CS	0.908
PSE -> EG	−0.583

Source Own elaboration with SmartPLS4 (Ringle et al. 2024)

Table 2.9 Specific indirect effects

Interactions among latent variables	Path coefficients
PSE -> EG -> CS	−0.792

Source Own elaboration with SmartPLS4 (Ringle et al. 2024)

Table 2.10 Total effects list

Interactions between latent variables	Total effects
CEP -> CS	0.494
EG -> CS	1.358
GBI -> CS	0.908
PSE -> CS	−0.792
PSE -> EG	−0.583

Source Own elaboration with SmartPLS4 (Ringle et al. 2024)

showing that the employment gap mediates the public expenditure in the comprehensive indicator related to context sustainability, that can represent that the public policies in terms of social expenditure related to reduce the employment gap are negatively related, in other words, the more is expended, the less the employment gap, contributing to context sustainability as a comprehensive indicator; it is important to consider that path coefficients are standardized regression coefficients, which must be interpreted as the change in the dependent variable if the independent variable is increased by one and all other independent variables remain constant (Ringle et al. 2024).

Next, for what it comes to the total effect list, these are useful for factor analysis, which has to be interpreted considering the level of significance of the p-value, as well as a confidence interval; for the analysis of the total effects, the results are included in the next Table 2.10.

To wrap out this section, the above table shows the total effects of the fitted model, showing an important positive effect from the Green bonds issuance to the context sustainability, as well as the employment gap and the carbon emission pricing; conversely, the public social expenditure shows a negative effect on context sustainability, as well as in the employment gap, which is an indication that the public policies associated in those American economies are not significant in solving employment gap by gender, as well as the overall context sustainability indicator.

Continuing with the analysis, the following figure shows the fitted model for the study; this represents the best model in terms of the quality criteria (R-square, CS, 0.982 and EG 0.340), (R-square adjusted, CS, 0. 971 and EG 0.246); and also the f-square (CEP -> CS 7.668; EG -> CS 45.596; GBI -> CS 29.786; PSE -> EG 0.515).

As it is possible to observe in Fig. 2.7, the fitted model shows significant positive effects from the Green bonds issuance (GBI) to the context sustainability (CS), as

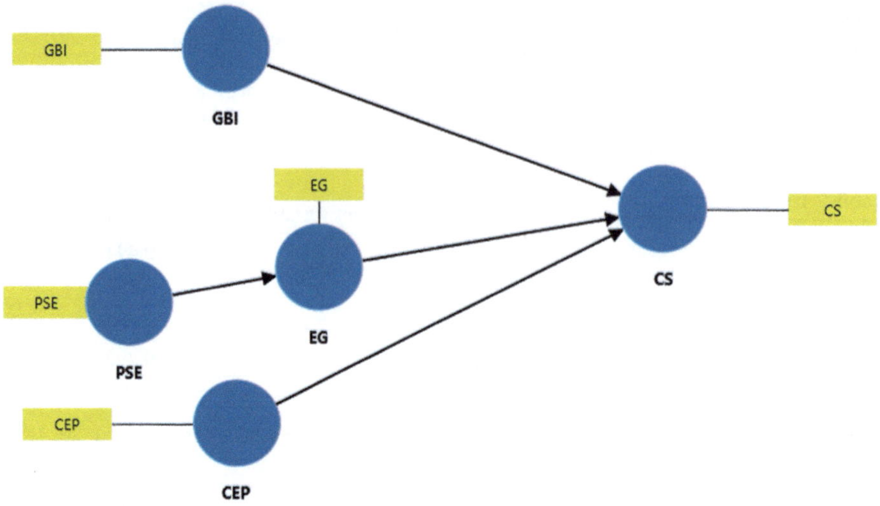

Fig. 2.7 Fitted SEM-PLS model for context sustainability. *Source* Own elaboration with SmartPLS4 (Ringle et al. 2024)

well as the employment gap (EG) and the carbon emission pricing (CEP); representing graphically the overall context sustainability indicator for the textile sector in America.

Conclusion

As the present research shows, the role of the government using the resources of the American economies in terms of designing and implementing public cash transfers (for Poorest and Richest Quintile) represents a statistically significant factor for sustainable practice in the textile industry sector, considering the products of the corresponding Chapter 58; this regards the relevance of the government as a salient stakeholder for this value chain.

In summary, for industries as well as for regions, to identify how sustainable practices contribute to competitiveness, a proper analysis will require the understanding of a comprehensive set of variables, which altogether are operationalized as interdependent constructs; in the case of a global value chain of the American economies included in the present research (Argentina, Brazil, Canada, Chile, Colombia, Ecuador, Mexico, Peru, and the United States of America), promoting sustainable practices requires the simultaneous creation of conditions in a set of relevant variables, such as promoting green bonds issuance (GBI), closing the employment gap (EG) and establishing a competitive carbon emission pricing (CEP) to achieve positive numbers in the context sustainability indicator.

References

Aguiar de Medeiros C, Trebat N (2017) Inequality and income distribution in global value chains. J Econ Issues 51(2):401–408

Ahmad N, Primi A (2017) From domestic to regional to global: factory Africa and factory latin America? In Chapter 3 in World Bank, pp 69–14. https://www.tandfonline.com/doi/full/, https://doi.org/10.1080/23322039.2021.2014654

Anner M (2021) Three labour governance mechanisms for addressing decent work deficits in global value chains. Int Labour Rev 160(4):611–629. https://doi.org/10.1111/ilr.12209

Anwar ST (2017) Zara vs. Uniqlo: leadership strategies in the competitive textile and apparel industry. Glob Bus Organ Excel 36(5):26–35. https://onlinelibrary.wiley.com/doi/https://doi.org/10.1002/joe.21805

Bartley T, Egels-Zandén N (2015) Responsibility and neglect in global production networks: the uneven significance of codes of conduct in Indonesian factories. Glob Netw 15(s1):S21-S44. https://onlinelibrary.wiley.com/doi/https://doi.org/10.1111/glob.12086

Beske P (2012) Dynamic capabilities and sustainable supply chain management. Int J Phys Distrib Logist Manag 42(4):372–387

Bruno FS, Filipecki ATP, Júnior ES (2009) Globa lisation of the textile sector and Brazilian preparation: search of the control of scarce assets of knowledge. Revista Espacios 30(1):22–34

Bunclark L, Barcellos-Paula L (2021) Sustainability reporting for sustainable supply chain management in Peru. Sustain Prod Consum 27:1458–1472. https://doi.org/10.1016/j.spc.2021.03.013

Dallas MP (2014) Cloth without a weaver: power, emergence and institutions across global value chains. Econ Soc 43(3):315–345. https://doi.org/10.1080/03085147.2014.881595

Evgeniev E, Gereffi G (2008) Textile and apparel firms in Turkey and Bulgaria: Exports, local upgrading and dependency. Econ Stud 17(3):148–179. https://doi.org/10.1080/00213624.2017.1320916

Frederick S, Gereffi G (2011) Upgrading and restructuring in the global apparel value chain: why China and Asia are outperforming Mexico and Central America. Int J Technol Learn, Innov Dev 4(1–3):67–95. https://www.inderscience.com/offers.php?id=41900

Frederick S, Cassill N (2009) Industry clusters and global value chains: analytical frameworks to study the new world of textiles. J Text Inst 100(8):668–681. https://doi.org/10.1080/00405000802137761

Garments RM (2014) Social sustainability in developing country suppliers: an exploratory study in the ready made garments industry of Bangladesh

Govindan K, Khodaverdi R, Jafarian A (2013) A fuzzy multi criteria approach for measuring sustainability performance of a supplier based on triple bottom line approach. J Clean Prod 47:345–354

Guang Shi V, Lenny Koh SC, Baldwin J, Cucchiella F (2012) Natural resource based green supply chain management. Supply Chain Manag Int J 17(1):54–67

Henseler J, Hubona G, Ray PA (2016) Using PLS path modeling in new technology research: updated guidelines. Ind Manag Data Syst 116(1):2–20. https://doi.org/10.1108/IMDS-09-2015-0382

Jones L, Demirkaya M, Bethmann E (2019) Global value chain analysis: concepts and approaches. J Int Com & Econ 1. https://heinonline.org/HOL/LandingPage?handle=hein.journals/jice2019&div=4&id=&page=

Le Breton-Miller I, Miller D (2006) Why do some family businesses out–compete? Governance, long–term orientations, and sustainable capability. Entrep Theory Pract 30(6):731–746

Leite MDP, Silva SRA, Guimarães PC (2017) The work in the clothing industry in São Paulo: the new forms of precariousness. Caderno CRH 30(79):51–67. https://www.scielo.br/j/ccrh/a/7N4ddYKK3JR95wBGBQz9tWM/abstract/?lang=en

Lu J (2015) Forecasting of US total textiles and apparel export to the world in next 10 years (2015–2025). J Text Appar, Technol Manag 9(2). https://www.semanticscholar.org/paper/Forecasting-of-U.S.-Total-Textiles-and-Apparel-to-Lu/ffdb1f8ce8a120f546c191eecb23892168bc7d0f

Markley MJ, Davis L (2007) Exploring future competitive advantage through sustainable supply chains. Int J Phys Distrib Logist Manag 37(9):763–774

Márquez R, Vásquez JA (2020) The extractivism of brown algae in northern Chile

Moreno-Brid JC, Gómez Tovar R, Sánchez Gómez J, Gómez Rodríguez L (2023) Trade and decent work in Mexico: the case of the automotive and textile industries. El trimestre económico 90(357):7–45. https://dialnet.unirioja.es/servlet/articulo?codigo=8796864

Ngo CN (2017) Industrial development, liberalisation and impacts of Vietnam–China border trade: the case of the Vietnamese textile and garment sector. Eur J East Asian Stud 16(1):154–184. https://brill.com/view/journals/ejea/16/1/article-p154_8.xml

Ozer SK, Taglioni D, Winkler D (2016) Turkey's participation and economic upgrading in global value chains. In Handbook of research on comparative economic development perspectives on Europe and the MENA region, pp 418–468. IGI Global. https://doi.org/10.1080/17543266.2011.577457

Ringle CM, Wende S, Becker J-M (2024) SmartPLS 4. Bönningstedt: SmartPLS. https://www.smartpls.com

Rodil-Marzábal Ó, Gómez Pérez AL, Campos-Romero H (2022) The global textile and apparel value chain: from Mexico–US–China linkages to a global approach. Economies 10(10):258. https://doi.org/10.3390/economies10100258

Sarkis J (2020) Supply chain sustainability: learning from the COVID-19 pandemic. Int J Oper Prod Manag 41(1):63–73

Tachizawa ME, Yew Wong C (2014) Towards a theory of multi-tier sustainable supply chains: a systematic literature review. Supply Chain Manag Int J 19(5/6):643–663

United Nations Environment Programme (2023) Sustainability and circularity in the textile value chain–a global roadmap. Paris. https://circulareconomy.europa.eu/platform/sites/default/files/2023-12/Full%20Report%20-%20UNEP%20Sustainability%20and%20Circularity%20in%20the%20Textile%20Value%20Chain%20A%20Global%20Roadmap_0.pdf

United States International Trade Commission (2011) Harmonized tariff schedule of the United States. United States International Trade Commission. https://www.wcoomd.org/~/media/wco/public/global/pdf/topics/nomenclature/instruments-and-tools/hs-nomenclature-older-edition/2007/hs-2007/1158_2007e.pdf?la=en

Van Zanten JA, Van Tulder R (2018) Multinational enterprises and the Sustainable development goals: an institutional approach to corporate engagement. J Int Bus Policy 1(3):208–233

Walker H, Jones N (2012) Sustainable supply chain management across the UK private sector. Supply Chain Manag Int J 17(1):15–28

Zhao L, Kim K (2021) Responding to the COVID-19 pandemic: practices and strategies of the global clothing and textile value chain. Cloth Text Res J 39(2):157–172

Chapter 3
Exploring Eco-Innovation Performance in the Textile Industry: Case of Two Export-Oriented Colombian Firms

Alex Jimenez-Diaz and Delia Echeverri-Vertel

Introduction

The textile and fashion industry is essential in the life of human beings and notable in the global economy. It manifests itself in 7% of global exports and employs more than 35 million people around the world (Desore and Narula 2018); however, it has generated a significant negative impact on the natural environment due to its high consumption of energy and water, as well as the generation of quantities of waste, derived from transformation processes (Moazzem et al. 2022), contributing to around 5% of total emissions (Sharma et al. 2021). It has been the second largest water and carbon footprint contributor, accounting for approximately 10% of global carbon emissions (Kasavan et al. 2021). According to Villemain (2019), fashion is the second-largest polluting industry in the world. It generates more gas emissions than all flights and maritime shipments, pours tons of microfibers into the oceans, and wastes large amounts of water in its production chain since manufacturing a single garment can lead to the use of 2700 L of water, equivalent to the consumption of an average person for two years (Molina 2019). In scenarios such as the European one, each individual, on average, consumes 26 kg of clothing per year and sheds 11 kg of textiles, reaching destinations similar to the Atacama Desert (Chile), which functions as a global dumping ground for clothing (Saez 2023).

To a large extent, these disproportionate situations are stimulated by high production of garments to meet the demands of global demand which, in the face of the current concern about negative impacts on the environment, global warming, and climate change, companies have led some efforts to adopt practices that minimize

A. Jimenez-Diaz (✉) · D. Echeverri-Vertel
University of Sinu, Montería, Colombia
e-mail: alexjimenezd@unisinu.edu.co

D. Echeverri-Vertel
e-mail: deliaecheverri@unisinu.edu.co

A. E. Toscano-Hernández et al. (eds.), *Sustainability, Stakeholders and Marketing in the Textile Sector*, SDGs and Textiles, https://doi.org/10.1007/978-981-97-6952-0_3

these impacts which require new types of products, processes, and organizational techniques. In this sense, eco-innovation (or green innovation, environmental innovation, ecological innovation) becomes an alternative that has attracted the attention of academics and practitioners, given the possibility of devising new means of generating competitive advantages for companies aligned with a growing environmental awareness of thousands of consumers around the world.

Some previous studies have addressed the incorporation of sustainability principles in textile production processes and the organization of the company, which constitute new development formats in the search to mitigate the negative impacts of the sector. Desore and Narula (2018) understand the strategies, barriers, and motivators that drive toward the implementation and promotion of responsible production and consumption actions in the textile sector. At the same time, Mac Lennan et al. (2021) identify the main initiatives of technological and non-technological eco-innovation of fashion companies in Brazil, from which derives a high prevalence of process innovation practices, followed by product innovation. In the same way, Rehman et al. (2021) pretended to understand the impact of consumers' environmental awareness on green performance in the textile industry; for this, they are concerned with inquiring about green culture, innovation, and commitment from the role of employees, who appreciate the position and perceptions of customers and constitute the ecosystem within the company.

Eco-innovation is a concept of environmental management that aims to eliminate the consequences of unsustainable business practices (Chen and Chen 2008). Mac Lennan et al. (2021) state that green innovation promotes establishing new markets in the face of the competitive dynamics of a green consumption model and applying restrictive environmental regulations in multiple markets. Chen et al. (2006) define it as innovation related to green products or processes to promote environmental management performance that satisfies environmental protection. It is partly motivated by the government and society's demand to contribute to sustainable development in different economic sectors.

In the textile industry, the performance of green innovation is dependent on the existence of resources, knowledge, experience, and capacities (Le Feber and Smit 2023) that, in the case of micro, small, and even medium-sized enterprises, are scarce, and they opt for cooperation activities between them. At the same time, Martínez-Martínez et al. (2023) determine that when organizations lack eco-innovation capabilities, their current or potential customers will be able to acquire them to transfer them to the company in an act of co-creation, which can be represented through the demands for products or services with a significant green component. Ahmed et al. (2023) in a study carried out in Pakistani textile companies, confirm that green innovation significantly impacts environmental and organizational performance and considers the role of organizational culture and managerial commitment relevant. Hence, many organizations in emerging markets consider corporate environmental practices challenging; however, they are convinced that they must adopt them and behave responsibly with the natural environment without losing economic focus (Sharma et al. 2021).

The Colombian textile industry is one of Latin America's oldest and most representative. According to the Single Business and Social Registry (RUES), Colombia has over 35,000 clothing companies, 200 yarn companies, and 7,000 textile companies. Of this register, less than 0.5% represent large companies (Ramirez-Rodriguez 2023), considered to be the ones that laid the foundations of a sector that later achieved great success. Its beginnings date back more than a century, as representatives Fabricato, Crystal, Lafayette, Encajes, S.A., and others such as Coltejer liquidated a couple of years ago. These companies, driven by government regulations and the demand of a customer with greater environmental awareness, have opted to implement environmental management programs that make them responsible companies with the natural environment and society. It also improves the corporate image and reputation, leading to a more significant positioning in the national and international markets.

This paper explores the performance of green innovation or eco-innovation in two large companies in the Colombian textile industry based on the analysis of sustainability reports and environmental practices in which the company has been venturing. To this end, it is decided to carry out a content analysis of official reports and web portals of the companies Fabricato and Crystal, as cases that transcend in Colombian history as the oldest and, in addition, pioneers in the adoption of environmental management and green innovation in their products, processes, and management activities. Likewise, the positioning of these two companies is consolidated both in the national and international markets, competing head-on with large companies worldwide.

This document, in addition to the introduction (section "Introduction"), is organized as follows. Section "Literature Review" defines the concepts of eco-innovation and types in the textile industry. Section "Method" on the method of work. Section "Results and Discussion" breaks down the findings of eco-innovation performance in companies in the Colombian textile industry. Finally, section "Conclusions" proposes some final reflections on green innovation in textile industry companies.

Literature Review

Eco-innovation or green innovation performance

Green innovation performance is defined by Chen et al. (2006) as "the performance of hardware and software involved in the innovation that a company carries out concerning green products or processes, including innovation in technologies, energy saving, pollution prevention, waste recycling, product designs, or corporate environmental management" (p. 332). This green innovation performance is divided into two parts. The first refers to *green product innovation* and is defined as the performance in product innovation related to the aspects mentioned above, and *green process innovation* corresponds to performance in the innovation process.

On the other hand, Hermundsdottir and Aspelund (2021), through a systematic review of 41 documents, identified several practices related to product innovation and green processes in the textile industry. This conceptual construction (Board 3.1) results in product innovation highlighting eco-design, eco-label, life cycle assessment, material, and packaging. Furthermore, they consider clean production, eco-efficiency, waste handling, supply chain management, and enzymatic textile processing in process innovation. They also include organizational innovation as an isolated component of process innovation, characterized by corporate policy, collaboration, business model innovation, knowledge culture and management, and risk management.

The scope of this paper addresses the assumptions of Chen et al. (2006). This perspective has been addressed by Singh et al. (2020) in their study on the interrelationship between transformational leadership, green human resource management, and green innovation, stating that green product and process innovation not only reduces negative environmental impact but also reduces the cost of waste and costs. In the same way, Xie et al. (2019) consider the innovation of green products and green processes, in which the first allows the modification or design of products considering the use of non-toxic components or biodegradable materials, and the second allows adjustments of business practice in the company's activities, in addition to having an impact on green product innovation.

In the textile sector, Sharma et al. (2021) consider only green product innovation as a construct to analyze green innovation; however, their study is extended

Board 3.1 Types of green innovation performance

Theme	Subtopics
Product innovation	Eco-design
	Eco
	Life cycle assessment
	Material
	Packaging
Process innovation	Clean production
	Eco-efficiency
	Waste handling
	Supply chain management
	Enzymatic textile processing
Organizational innovation	Corporate policy
	Collaboration
	Business model innovation
	Culture and knowledge management
	Risk management

Note Taken from the work of Hermundsdottir and Aspelund (2021)

to green performance and managerial commitment, which employs green process actions. While Ahmed et al. (2023) define green innovation as services, products, and processes aimed at reducing the degradation of the natural environment and making it friendlier, they specify that product innovation and process innovation are the two dimensions of green innovation. Finally, Salo et al. (2020) state that green innovation is dimensioned through green product and process innovation. The first dimension focuses on the choice of the type of recycled material with lower emissions, as well as the implications for redesigning the product with fewer materials. On the other hand, the second focuses on reducing the use of raw materials, water, electricity, and the waste generated in the production process. In this way, the concepts of green innovation coincide with those proposed by Chen et al. (2006), therefore, for this study, only the concepts of green product innovation and green process innovation will be considered, considering the characteristics provided by Hermundsdottir and Aspelund (2021) (Board 3.1).

Method

This research is of a documentary type, characterized by being a process based on the search, retrieval, analysis, critique, and interpretation of secondary data (Arias 2012) taken from the official web portals of the companies studied. Similarly, this work is simplified into the study of two cases of companies in the Colombian textile industry, Fabricato S.A. and Crystal S.A.S., considering the case study as a methodology that provides an adequate framework for the understanding of the unit of analysis and the context (Yin 2009). The documentary review is based on the content analysis of sustainability reports and reports of the companies studied, which allows for describing and making reproducible and valid inferences from the data and the context (Krippendorff and Wolfson 1990), which recreates an idea of what exists and that can be replicated, in academic and business terms.

For this work, Fabricato and Crystal were considered two Colombian companies in the textile sector, leaders in the Colombian market in the supply of fabrics for clothing and other activities.

Results and Discussion

Fabricato (www.fabricato.com)

Fabricato is a Colombian company dedicated to manufacturing and marketing textile products, which has existed for more than 100 years. Today, this company is based on a vertically structured business model and oriented on four approaches (fashion brands, endowment market, textile supply, and products for specialized application) with a transversal commitment to sustainability that consists of an investment in environmental responsibility that exceeds 9,000 million Colombian pesos at the end of 2022. A significant figure is represented in the actions that the company has been taking on track.

Crystal (www.crystal.com.co)

Crystal is a Colombian company, producer, and marketer of clothing and fashion brands under spinning, textile supply, and own brands, created more than 90 years ago. It has accelerated growth, derived from business diversification strategies, as well as the expansion of its production chain, acquisition of its stores, and export enhancement, allowing it to have more than 270 stores in Colombia, 6 manufacturing plants, and participation in 11 Latin American countries today. This company is no stranger to changes in the environment and recognizes that the adoption of sustainable initiatives not only leads it to have a significant impact on society but also to reduce the high negative impact that the textile industry provides to the environment; therefore, its business model adopts sustainable innovation, which incorporates aspects of green innovation.

Analysis

The content analysis determined three forms of green innovation in the textile industry, which are linked to the dimensions proposed by Chen et al. (2006): eco-design and ecological material (green product, eco-packaging, and eco-label), eco-efficiency and clean production, and environmental management. In this way, green product innovation is related to ecological design (eco-design) and eco-friendly materials. On the other hand, green process innovation is evidenced by eco-efficiency, clean production, and environmental management practices.

Eco-design and eco-friendly material

Ecodesign is determined as a way to integrate environmental aspects and product and process development (Salo et al. 2020). According to the 2023 sustainability report (Fabricato 2023), Fabricato is concerned with generating quality of life for nature based on the design and creation of products with ecological materials, recovered fibers, and originating from alternative research such as the leaf of the pineapple plant, constituting eco-design and the production of products as a dependent link in the traceability of the production and value chain. For its part, according to the 2022 sustainability report (Crystal 2022), the textile industry company Crystal addresses the production of environmentally friendly products by involving inputs (such as cotton) grown from areas constituted as responsible; likewise, it is aware that the process of making garments requires the use of dyes; consequently, it is committed to reducing water consumption and the use of biodegradable inputs.

Both Fabricato and Crystal S.A.S. ensure an eco-friendly product that also considers the implementation of reverse supply chain systems, in which labels and packaging adhere to reuse and transformation processes with 100% recyclable inputs or produced from materials based on sugarcane bagasse, as an alternative to reduce water and energy consumption in the production process.

Eco-efficiency and clean production

Eco-efficiency is used to stimulate productivity and innovation and improve environmental performance. Bidoki et al. (2006) propose that eco-efficiency studies the product from the extraction of the raw material until after it has been used, following three critical objectives: (1) Increase the value of the product or service, (2) Optimize the use of resources, and (3) Reduce environmental impact. On the other hand, associated with the concept of eco-efficiency is clean production, which, according to de Oliveira Neto et al. (2019), is defined as the integration of the application of environmental strategies to products, processes, and services to increase efficiency and reduce risk to the environment and society.

In this sense, eco-efficiency and clean production are similar and conducive to ordinary activities in practice. In the case of companies in the textile industry, especially those studied in this work, they make efforts to minimize the consumption of essential resources, such as water and energy, through the implementation of treatment plants and the incorporation of alternative energies, as well as the adoption of programs aimed at reducing CO_2 emissions. These companies use eco-efficient machinery and computer equipment with lower energy consumption, and they use high-tech LED lighting, which is consistent with their environmental policies.

Environmental management

For organizations to operationalize environmental practices, they must have a model of thinking channeled through policies from corporate governance. So, environmental management leads to the integration of environmental aspects in the management of the company (Hermundsdottir and Aspelund 2021), making an adequate distribution and efficiency of items for investment in environmental systems, the structuring of an approach that includes an environmental strategy with medium and long-term achievements. In the case of the companies under study, they have strengthened their ties with different actors in the supply chain, delivering eco-friendly products to their customers and improving the acquisition of inputs from their suppliers.

In addition, these companies have been recognized by environmental associations and have committed to improving processes that, consequently, have resulted in certifications that attest to the actions incurred by the company to improve landfill management. Likewise, they have encouraged their strategic allies to incorporate green practices in their organizations as a strategy for environmental conservation of the business ecosystem, especially in an industry classified as highly polluting (Board 3.2).

Board 3.2 Green innovation in Colombian companies in the textile industry

Green innovation	Fabricato	Crystal
Eco-design and eco-friendly material (product, eco-packaging and eco-label)	• Products made from polyester recovered from PET bottles • Packaging bags are produced with recovered material • Product lines (denim, knit, denim, and poplin) incorporate recovered fibers • Denim product lines with technology with lower water consumption • Extraction of fibers from leaves of the pineapple plant • Ecolabelled under ISO 14021-I standard (sustainable dumping practices and zero discharges of hazardous chemicals)	• Production of garments with certified cotton grown in responsible conditions • Denim is produced with biodegradable dyes and reduces up to 40% of water used for washing • Production of garments with sustainable cotton • Use of Lyocell, EcoVero Viscose, recycled fibers, and raw materials that are part of eco-responsible production processes and protect the earth • Garment packaging in cardboard without laminates or glues • Reusable bag (packaging) made with recycled fabric from PET bottles • Labels are recyclable and biodegradable and made with sugarcane bagasse

(continued)

Board 3.2 (continued)

Green innovation	Fabricato	Crystal
Eco-efficiency (and clean production)	• Trigeneration of energy (electrical, thermal, and gases) • Reduction in using virgin resources • 92% reduction in water consumption of indigo dye for the Denim line • Eco products incorporating recovered fiber, technologies that reduce the demand for resources, and polyester from PET bottles • Programs for the efficient use of water, air, Soil, Energy and Biodiversity • Wastewater treatment plant for industrial reuse (65% of used water is reused) • Reduction of textile waste and landfill occupation from using packaging inputs derived from by-products • 94% of self-generated energy and 71% of energy consumed from renewable sources	• Use of sustainable wood. Drywall construction material (reduces water consumption and waste generated) • Reduction in the use of water resources • Use of organic cotton from crops free of chemical fertilizers and pesticides. 90% less water consumption and 60% less energy. Biodegradable and recyclable • 89% energy efficiency in computer equipment • High-tech LED lighting • 27% reduction in energy consumption • Savings in clothing with 3D modeling • Use of solar energy in own production plants and those of strategic allies • Water treatment plant • Use of the paper bag with a single water-based ink
Environmental management	• OEKO-TEK international certificate standard 100 (toxic-free chemical products) • Participation in associations and agreements for environmental performance • Adequate conservation of 280,000 trees in self-protected areas • Reduction of shipments to landfills • Biodegradable cups produced with sugarcane bagasse fibers are used • Electric vehicle fleet to reduce CO_2 emissions • Area in plant destined for recovery, Commercialization of a line of fabrics made with fibers recovered from textile processes and customer clothing scraps	• Certification in responsible and safe production • Delivery of products from strategic allies with the use of recycled tires and biodiesel to reduce CO_2 emissions • Adoption of a circular economy model in the reuse of hooks • Declares principles to be a sustainable company by creating new ideas and constantly changing the value chain

Note Elaboration based on the sustainability reports generated by Fabricato (2023) and Crystal (2022)

Conclusions

The textile industry, historically rooted in processes that consume high water, energy, and harmful chemicals, faces an urgent need for change toward environmental sustainability. Implementing innovative models focused on a more friendly ecosystem with the natural environment opens the door to the creation and application of alternatives that minimize its negative impact. That is why, through improvements in techniques and materials in the production of products, it is possible not only to reduce the ecological footprint but also to improve the efficiency in the use of resources and reduce the dependence on substances harmful to both the environment and human health.

This shift toward green innovation in the textile industry implies a transformation in production processes and a change in mindset toward more sustainable practices. At the same time, the implementation of recycling and reuse systems, as well as the promotion of transparency in the supply chain, are vital aspects of moving toward a more ethical and responsible business model. In addition, this transition to sustainability benefits the environment and can lead to economic opportunities, such as creating new markets for eco-friendly products and improving the reputation of brands committed to protecting the planet.

The performance of green innovation (or eco-innovation) in the textile industry is qualified as a mechanism conducive not only to reducing risks in the environment but is also used as a strategy that improves the corporate image, generating representative effects on the economic performance of organizations.

This study aims to determine that the green innovation seen in companies in the Colombian textile industry is channeled through eco-design and ecological material comprising green product innovation, eco-efficiency, clean production, and environmental management, which analyze green process innovation.

References

Ahmed RR, Akbar W, Aijaz M, Channar ZA, Ahmed F, Parmar V (2023) The role of green innovation on environmental and organizational performance: moderation of human resource practices and management commitment. Heliyon 9(1):e12679. https://doi.org/10.1016/j.heliyon.2022.e12679

Arias FG (2012) El proyecto de investigaci6n. Introducci6n a la metodología científica. 6ta edici6n. Editorial Episteme

Bidoki SM, Wittlinger R, Alamdar AA, Burger J (2006) Eco-efficiency analysis of textile coating materials. J Iran Chem Soc 3(4):351–359. https://doi.org/10.1007/BF03245958

Chen Y-S, Lai S-B, Wen C-T (2006) The influence of green innovation performance on corporate advantage in Taiwan. J Bus Ethics 67(4):331–339. https://doi.org/10.1007/s10551-006-9025-5

Chen Y, Chen Y (2008) The driver of green innovation and green image–green core competence. J Bus Ethics 531–543. https://doi.org/10.1007/s10551-007-9522-1

Crystal SAS (2022) Informe de sostenibilidad 2022. Somos Crystal. https://www.crystal.com.co/sustainabilityReports

de Oliveira Neto GC, Ferreira Correia JM, Silva PC, de Oliveira Sanches AG, Lucato WC (2019) Cleaner production in the textile industry and its relationship to sustainable development goals. J Clean Prod 228:1514–1525. https://doi.org/10.1016/j.jclepro.2019.04.334

Desore A, Narula SA (2018) An overview on corporate response towards sustainability issues in textile industry. Environ Dev Sustain 20(4):1439–1459. https://doi.org/10.1007/s10668-017-9949-1

Fabricato SA (2023) Informe de sostenibilidad ambiental y social 2023. https://site.fabricato.com/sostenibilidad/

Hermundsdottir F, Aspelund A (2021) Sustainability innovations and firm competitiveness: a review. J Clean Prod 280. https://doi.org/10.1016/j.jclepro.2020.124715. (Elsevier Ltd.)

Kasavan S, Yusoff S, Guan NC, Zaman NSK, Fakri MFR (2021) Global trends of textile waste research from 2005 to 2020 using bibliometric analysis. Environ Sci Pollut Res 28(33):44780–44794. https://doi.org/10.1007/s11356-021-15303-5

Krippendorff K, Wolfson L (1990) Metodología de análisis de contenido: teoría y práctica. Paidos

Le Feber N, Smit MJ (2023) Fashion companies pioneering with eco-innovations in the swedish fashion industry: motivations, resources, and cooperation. Circ Econ Sustain 3(4):1885–1905. https://doi.org/10.1007/s43615-022-00246-x

Mac Lennan MLF, Tiago EF, Pereira CEC (2021) Technological and non-technological trends in fashion eco-innovations. Innov Manag Rev. https://doi.org/10.1108/INMR-12-2020-0177

Martínez-Martínez A, Cegarra-Navarro J-G, Garcia-Perez A, De Valon T (2023) Active listening to customers: eco-innovation through value co-creation in the textile industry. J Knowl Manag 27(7):1810–1829. https://doi.org/10.1108/JKM-04-2022-0309

Moazzem S, Crossin E, Daver F, Wang L (2022) Environmental impact of apparel supply chain and textile products. Environ Dev Sustain 24(8):9757–9775. https://doi.org/10.1007/s10668-021-01873-4

Molina P (2019, Diciembre 16) ¿Cuántos litros de agua llevamos puestos? Iagua. https://www.iagua.es/blogs/paz-molina-elias/cuantos-litros-agua-llevamos-puestos-0

Ramirez-Rodriguez JC (2023) Tratamiento de aguas residuales y problemáticas ambientales del sector textil en Colombia: una revisión. Informador Técnico 87(1):82–106. https://doi.org/10.23850/22565035.5304

Rehman SU, Kraus S, Shah SA, Khanin D, Mahto RV (2021) Analyzing the relationship between green innovation and environmental performance in large manufacturing firms. Technol Forecast Soc Change 163:120481. https://doi.org/10.1016/j.techfore.2020.120481

Saez C (2023, January 30) ¿Sabemos cuánta ropa se recicla y cómo contamina? Revista Técnica de Medio Ambiente (RETEMA). https://www.retema.es/actualidad/sabemos-cuanta-ropa-se-recicla-y-como-contamina

Salo HH, Suikkanen J, Nissinen A (2020) Eco-innovation motivations and ecodesign tool implementation in companies in the Nordic textile and information technology sectors. Bus Strat Environ 29(6):2654–2667. https://doi.org/10.1002/bse.2527

Sharma S, Prakash G, Kumar A, Mussada EK, Antony J, Luthra S (2021) Analysing the relationship of adaption of green culture, innovation, green performance for achieving sustainability: Mediating role of employee commitment. J Clean Prod 303:127039. https://doi.org/10.1016/j.jclepro.2021.127039

Singh SK, Giudice MD, Chierici R, Graziano D (2020) Green innovation and environmental performance: the role of green transformational leadership and green human resource management. Technol Forecast Soc Change 150. https://doi.org/10.1016/j.techfore.2019.119762

Villemain C (2019, abril 12) El costo ambiental de estar a la moda. Programa de Las Naciones Unidas Para El Medio Ambiente. https://news.un.org/es/story/2019/04/1454161

Xie X, Huo J, Zou H (2019) Green process innovation, green product innovation, and corporate financial performance: a content analysis method. J Bus Res 101:697–706. https://doi.org/10.1016/j.jbusres.2019.01.010

Yin RK (2009) Case study research: design and methods, vol 5. Sage

Chapter 4
Transformational and Sustainable Leadership in Textile Companies in Lima: A Qualitative Study

Eli Malvaceda-Espinoza, Pierina Cubillas-Espinoza, Gianpiero Flores-Pereyra, and Fiorella Rivera-Cruzatt

Introduction

During 2020, the pandemic caused by COVID-19 generated a global economic crisis that especially affected small, medium and micro enterprises (World Bank 2023a). Even today, Latin American and Caribbean countries are recovering, especially in terms of income and employment levels (World Bank 2022). In Peru, in 2022, the economic recovery was manifested in the growth of indicators such as the Gross Domestic Product (GDP) (+2.7%), boosted by the primary and services sectors (World Bank 2023b). In addition, the number of companies in Peru has been increasing, reaching 3.25 million in the second quarter of 2023, with the main sectors being: commerce, business services, food and manufacturing (Instituto Nacional de Estadística e Informática (INEI) 2023). In light of these changes, businesses globally are prompted to rethink their strategies and leadership approaches. The relevance of addressing these issues aligns with the United Nations Sustainable Development Goals (SDGs), which promote global economic, social, and environmental sustainability (United Nations 2024a).

E. Malvaceda-Espinoza (✉) · P. Cubillas-Espinoza · G. Flores-Pereyra · F. Rivera-Cruzatt
San Ignacio de Loyola University, Lima, Peru
e-mail: emalvaceda@usil.edu.pe

P. Cubillas-Espinoza
e-mail: pierina.cubillas@usil.pe

G. Flores-Pereyra
e-mail: gianpiero.flores@usil.pe

F. Rivera-Cruzatt
e-mail: fiorella.riverac@usil.pe

A. E. Toscano-Hernández et al. (eds.), *Sustainability, Stakeholders and Marketing in the Textile Sector*, SDGs and Textiles, https://doi.org/10.1007/978-981-97-6952-0_4

While these transformations are unfolding, the Peruvian textile sector, which is made up predominantly of small apparel companies (90%), is facing various difficulties such as obtaining financing, training, and managing a larger workload (Del Carpio 2023), magnifying the necessity for a strategic focus on leadership. In this scenario, competitiveness is necessary to stand out among the more than 46 thousand companies that exist in the textile sector, where nearly 400 thousand jobs are created directly and 300 thousand jobs indirectly (Tello 2019).

In turn, pandemic triggers for organizations were evidenced when workers experienced instability between work and their daily lives, as the way in which and where employees work changed during teleworking, leaving employees without sufficient information, guidance, and expected results about the work they have to do (Charoensukmongkol and Puyod 2021; Prasad et al. 2020), highlighting the need for more effective leadership. In view of this, there is a need to know the aspects that allow the achievement and maintenance of business objectives, such as leadership styles (Jiménez and Villanueva 2018) and the impact of organizational culture (Palafox-Soto et al. 2021). This understanding is crucial for ensuring the productivity and resilience of companies not only in the face of immediate challenges but also in laying the groundwork for sustainable productivity and development in the future.

In this line, leaders should seek sustainable development in their companies, whether this is in environmental, economic, and/or social terms (Correia 2019). Based on this, the following question is posed: How is transformational and sustainable leadership experienced by managers of textile companies in Lima?

Literature Review

Both transformational and sustainable leadership are essential to address current and future challenges. In this regard, transformational leadership, which seeks to motivate followers to drive positive change, aligns with the principles of the SDGs, such as decent work and economic growth (SDG 8) (United Nations 2024b), reduction of inequalities (SDG 10) (United Nations 2024c) and responsible consumption and production (SDG 12) (United Nations 2024d).

To comprehend the role of leadership in achieving sustainability goals, it is crucial to recognize leadership as the ability to influence and guide group members toward the achievement of shared objectives. In this sense, the leadership style of managers will have a significant impact on business productivity, as it focuses on structuring and coordinating resources, as well as maintaining the stability of the company (Purwanto 2021; Van Diggele et al. 2020).

This is how the style of transformational leadership stands out, as it implies that the leader motivates their followers to foster positive changes, not only through material incentives but also by contributing to their personal and professional growth (Mendoza et al. 2014; Bass and Avolio 2006). Moreover, the type of leadership in question is based on four main areas (Bass 1985): individual consideration, which is manifested when they devote attention to the personal and professional development of the team, inspirational motivation, in which leaders generate motivation in their followers through inspiration and a meaningful emotional connection; intellectual stimulation, from where creativity and critical thinking are promoted in the team; idealized influence, in which the leader receives respect and admiration through charisma (Jauregui-Arroyo et al. 2022; Muralidharan and Pathak 2018).

Leaders who are both transformative and sustainable effectively communicate the company's sustainable vision through a coherent overview, addressing issues related to the strategies necessary for achieving sustainable goals and the methods involved in implementing these strategies. In this regard, leaders with this type of leadership can instill self-confidence, optimism, happiness, and enthusiasm in employees by engaging in regular discussions, as receiving support reduces stress, burnout, and depression (Tang et al. 2022).

Thus, these leaders promote the well-being of employees by modifying their perceptions of work and fostering a sense of community, encouraging followers to exhibit a commitment higher than expected, urging them to provide support, motivating them to seek continuous improvement, and providing the necessary resources to achieve their goals. In this way, the expressed job satisfaction becomes permanent and favorable (Burke and Richardsen 2019; Purwanto 2021). As well, there is a difference between female and male leaders, as the former exhibit higher levels of transformational leadership behaviors (individual consideration) and lower levels of passive leadership behavior (Bass 1998).

It is worth mentioning that this leadership should seek to be sustainable; that is, it should be focused both on improving the quality of life of workers, as well as generating income (for the present and future) for the organization (Iqbal and Ahmad 2021). In addition, it promotes capacity building, sustainable change, and long-term results. Among its practices are valuing collaborators, shared vision, having friendly labor relations and having social responsibility, and improving the long-term performance of organizations (Iqbal and Ahmad 2021). Therefore, the practices and strategies adopted for leaders' decision-making will promote sustainable working, and social and human conditions (Lämsä et al. 2023).

In this regard, Gil-Gómez et al. (2020) considered that corporate sustainability is based on meeting the present needs of the company without compromising the ability of future generations to meet their own needs. In turn, it includes economic sustainability, by seeking to achieve a balance between the pursuit of economic performance and its sustainable development. It also includes environmental sustainability, which refers to the influence of an organization's business processes, activities, and operations on other entities. Finally, it encompasses social sustainability, as it promotes a partnership and acceptance between business and society (Gil-Gómez et al. 2020; Nosratabadi et al. 2019).

Among the studies conducted in European and Asian countries such as Turkey, Holland, Belgium, China, France (Erden and Yaşlioğlu 2020), Indonesia (Evitha et al. 2022), Pakistan (Ribeiro et al. 2018), China (Saeed and Jun 2021) and a systematic study (Hyun-Duck and Bautista 2022), it was found that this type of leadership has as a fundamental pillar the social support and valuing toward the members that compose the work team through recognition and praise. Studies report that leader-driven motivation promotes individual performance (Erden and Yaşlioğlu 2020; Ribeiro et al. 2018), thus, allowing to enhance the worker's capabilities (Erden and Yaşlioğlu 2020), particularly when such leader has experience and skills to set an example to others (Hyun-Duck and Bautista 2022; Saeed and Jun 2021), in which gender is one of the sociodemographic characteristics that will influence the form of leadership (Hyun-Duck and Bautista 2022).

Regarding studies about sustainable leadership reported in countries such as Thailand (Burawat 2019), Ghana (Nsiah-Sarfo et al. 2023) and 25 other countries (Leal et al. 2020), it was found that technical capabilities for innovation, complexity management, and long-term vision are elemental to engage staff. It is reported that constant training and communication are promoters of the fulfillment of objectives. In addition, the systematic study by Cantón-Mayo et al. (2021) concludes that emotional components, such as passion, trust, expectations, imagination, integrity, commitment, curiosity and openness, stress and interpersonal relationship management, and leader support are important to promote sustainable leadership.

The background review identified an absence of research that takes into account transformational and sustainable leadership in the context of textile companies. Likewise, there is a lack of research on a qualitative approach to analyze leadership. Moreover, the research mentioned are from international contexts, which highlights the need for studies in the Peruvian context. In view of this, the study proposes to learn about the experiences of transformational and sustainable leadership from managers of textile companies in Lima.

Methodology

The research followed a qualitative approach that allowed us to delve deeper into the experiences of individuals around transformational and sustainable leadership (Creswell and Poth 2018). In addition, a thematic analysis design followed, which allows us to identify, analyze, and understand patterns based on the relevant themes expressed by the participants empirically, organized through codes and themes of meaning (Braun and Clarke 2022).

Participants

There were six managers in textile companies in Lima. A convenience sampling was used due to the limitations of access to the participants. Inclusion criteria were considered to be between 38 and 55 years of age, with a university or technical education. Participants were excluded if they had any illness at the time of data collection.

Table 4.1 Participants' characteristics: managers in the textile industry

Pseudonym	Gender	Age	Level of education	Management area
JV	Male	38	Bachelor	Commercial
CV	Male	53	Technical	General
MO	Female	49	Technical	Administrative
FB	Male	47	Bachelor	Sales
EM	Female	55	Bachelor	Administrative and finance
FBP	Male	53	Technical	Commercial

Note The participants come from different companies in the textile industry

All the companies were dedicated to the manufacture of garments, were founded between 5 and 10 years ago, and had their main headquarters in metropolitan Lima. All of them are located in the category of medium-sized companies (which have annual sales greater than between 1,700 and 2,300 tax units) (Congress of the Republic of Peru 2013). Regarding the characteristics of the participants, there were two women and four men. The age range was between 38 and 55 years (average 49). In terms of education levels, three had university studies and the other three had technical studies. All are managers in different areas: Commercial, Sales, Administration and Finance. The characteristics of the participants are presented in Table 4.1.

The participants are mostly male, adult, commercial, and administrative managers, and in terms of educational level, they are evenly divided between those with technical and university studies. This is consistent with the traditional profile of managers in metropolitan Lima. No relevant differences were found in the subgroups indicated.

Instruments

For the present study, a semi-structured interview based on the theoretical contributions of Bass (1985) in relation to transformational leadership, and of Cantón-Mayo et al. (2021) in relation to sustainable leadership were chosen. A categorization matrix was constructed in which the following topics were identified: individual consideration, intellectual stimulation, inspirational motivation, idealized influence, psychological tolerance, as well as technical capabilities, and emotional components. Likewise, a topic referring to familiarization questions was added, in which sociodemographic aspects of the participants were asked: type of company, sex, age, level of education, and position. The assessment of the instrument's rigor begins with the development of a categorization matrix based on the aforementioned theoretical contributions. Additionally, an interview guide was prepared and reviewed by (3) specialists in the field prior to its implementation. Subsequently, adjustments were made to the content pertaining to questions related to economic sustainability until the interview protocol was ready for the pilot test, which was conducted with a participant possessing characteristics similar to those of the study (Table 4.2).

Table 4.2 Topics and questions for the interview guide

Topics	Interview questions
Individual consideration	How do you act when a problem arises with the personnel you are in charge of, and in what position do you place yourself to try to resolve the situation?
Intellectual stimulation	How do you promote the development of skills in your staff, how do you encourage creativity in your team?
Inspirational Motivation	How do you motivate your staff to solve their problems? What perspective do you take to face a problem with your team?
Idealized influence	How do you express your confidence in your staff, how much consideration do you give to the interests of your personnel?
Psychological tolerance	How do you handle interpersonal problems with your staff, how do you handle problems that arise between them?
Technical capabilities	When a member of your staff has a difficulty, how do you act, how do you give feedback to your staff, and what kind of answers do you look for from them?
Emotional components	How do you communicate the vision of your area to the personnel you are responsible for, how do you manage respect with the people you are responsible for?

The table presented is useful in the research because it allows us to highlight the topics to be addressed in the interview guide. It is also important to note that the questions are oriented to respond to the research objective and are also contextualized according to the participants.

Procedure

Authorization was requested from the human resources area of the different companies, and once this was approved, those responsible for this area were asked to contact people who met the profile of the participants. Subsequently, contact was made with the potential participants, who were informed of the research objective and the procedures to be followed in the research. To those who agreed to participate voluntarily, the informed consent form was explained, reviewed, and signed. After that, the dates and times of the interviews were coordinated. The interviews were face-to-face and took place in a private place within each company. The interview was conducted with a single interviewer who was a member of the research team. The interviews were audio-recorded and lasted an average of 40 min. There were no people who refused or dropped out of the study. No incentive or economic benefit was given for participation in the research. None of the researchers were part of the companies at the time of the interviews. The interviews were conducted between January and February 2020.

Ethical and Quality Aspects

The principles outlined in the Helsinki Declaration were followed, as well as the principles of professional responsibility of the Colegio De Psicólogos del Perú (Consejo Directivo Nacional 2017). The principles of autonomy, confidentiality, anonymity, beneficence, nonmaleficence, and fairness to the participants were assumed. Regarding the criteria of rigor, the study sought to cover the criterion of credibility through the return of the transcripts and results to the participants, so that they could review and evaluate what was found in the study. Likewise, the auditability criterion was sought to be ensured, since the interview protocol, as well as the research, was evaluated by an expert in qualitative methodology during the process. Finally, the criterion of dependence was met, since the data analysis was conducted as a team to minimize bias at the time of coding (Daniel 2018).

Data Analysis

Hybrid thematic analysis (Swain 2018) was used for information analysis. It was based on a theoretical approach to the construct for the generation of a priori codes. Subsequently, familiarization with the information collected and transcribed literally was carried out, this involved reading the texts, as well as the elaboration of quotations (sections of text that can be phrases, sentences, or sets of sentences) that respond to the research objective. Next, the initial codes were created using deductive (by means of the aprioristic codes) and inductive coding, initiating the first coding cycle in which open, descriptive, and thematic coding was carried out.

Subsequently, the codes were refined and grouped, identifying thematic patterns. At this point, the second coding cycle was applied and selective, axial and pattern coding was used. Subsequently, the definition of themes was carried out by reviewing the co-occurrence of codes to identify the relationships between them and thus be able to represent them in a semantic map. Finally, the report was elaborated taking into account the significant codes, which were obtained through the tactics of meaning generation, these being representativeness, frequency, and density (Miles et al. 2014). The analysis software used was ATLAS.ti 24 and performed by the research team.

Results

After analyzing the documents, the experiences of transformational and sustainable leadership from managers of textile companies were identified, highlighting four themes: individualized leadership, inspirational stimulus, learning based on mistakes and experiences, and exemplary authority (Fig. 4.1).

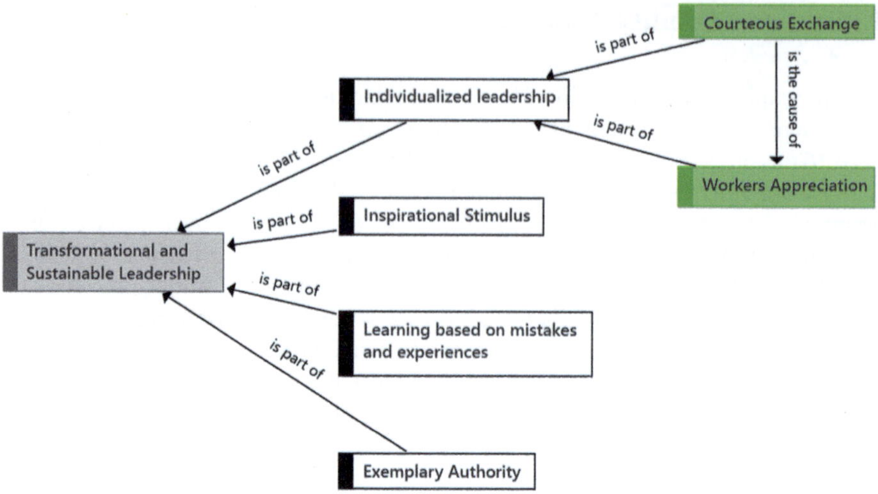

Fig. 4.1 Semantic map of transformational and sustainable leadership. *Source* Own elaboration, with the support of ATLAS.ti 24

The four main themes found around transformational and sustainable leadership in textile companies in metropolitan Lima can be seen in Fig. 4.1. Also, it is indicated that individualized leadership, in turn, includes the courteous exchange and workers appreciation. All the relationships identified are of transitive belonging ("is part of") and were determined from the visualization of the co-occurrences between the codes.

Individualized Leadership

The participants noted that a leader with these characteristics is concerned with attending to the needs of each employee and promoting the development of his or her skills in an individualized manner. They also stressed the importance of establishing a *courteous exchange* between the leader and the employee, based on active listening, trust, empathy, and mutual respect. As well, they also emphasized the support and backing they provide in both work and personal matters. "Being friendly, empathetic, good-natured and trying to transmit positive energy in my environment [is the way I transmit confidence to the staff]" (MO, 49 years old, Female, Administrative Manager).

Managers also reported keeping in touch with their staff, giving them feedback through advice. "I talk to him; I keep track to see what he thinks or how he feels. (EM, 55 years old, Female, Manager of Administration and Finance). They also provide a space for conversation "I always strive to maintain closeness with them, offering advice and facilitating their overall improvement (…) if they require a space for conversation, I readily provide it" (JV, 38 years old, Female, Commercial Manager).

Finally, they take into account the needs of workers, even when these are personal. "When my staff have a problem, I try to support them. If they ask for permission, I give it. Sometimes I am a bit condescending on these kinds of issues." (FBP, 53 years old, Male, Commercial Manager).

They indicated that the way to consider each employee is by *valuing their capabilities*, which implies trusting in their abilities to solve challenges or difficulties by their own means. "I always encourage them to look for solutions, to ask questions and read if necessary. I trust them to be able to face challenges successfully" (MS, 55 years old, Female, Administrative and Financial Manager). They also mentioned the importance of learning from staff and asking their opinion to solve problems. "I learn from my collaborators, even if I have more experience, you never stop learning." (JV, 38 years old, Female, Commercial Manager).

> I think this is something I should do more (make staff perceive difficulties as learning). Sometimes I stress it, but most of the time they figure it out on their own [...] You keep learning from the people you work with... just like they can learn from me, they teach me some things, especially on technology issues [laughs] (CV, 53 years old, Male, General Manager)

They even highlight the positive results they get because of their own learning. "We have meetings from time to time, and that's when important topics come up. I take this opportunity to highlight the good results we have obtained." (FBP, 53 years old, male, commercial manager). These narratives were told mainly by male participants.

In synthesis, the polite exchange highlights the forms of empathetic communication and the mutual and individual support of the leader with the collaborators, fostering a relationship of trust and openness. At the same time, it brings as a consequence the valuation of the collaborators, highlighting the recognition of the personnel's capabilities and the understanding of the complexities that may arise in their lives. It is worth mentioning that these two elements, in turn, are part of the individualized leadership.

Inspirational Stimulus

Is characterized by inspiring followers and establishing a deep emotional connection to *stimulate motivation.* In that sense, they expressed an inspiring vision, motivating their followers, generating high expectations, and fostering confidence in their goals. In addition, they were willing to address shared objectives and emphasized the importance of communicating the vision. "(I convey the vision) always trying to do things as well as possible, analysing situations or processes to make them more effective and simple which I make known." (MO, 49 years old, Female, Administrative Manager).

Likewise, they communicate the area's plans, so that employees recognize the company's vision and generate a joint commitment, even communicating the plans they have so that workers feel integrated, both with the leader's work and with the company. "I will outline my plans, with a focus primarily on the company's direction.

It is crucial to involve your staff, so they feel integral to the company; I believe this fosters their commitment to our objectives" (JV, 38 years old, Female, Commercial Manager).

In the same way, the participants mentioned that sometimes it is a challenge to be able to communicate the company's vision, which involves a challenge for leadership. "I try to do it (convey the vision by being an example), but sometimes it is a challenge. Ideally, I would like to be consistent, but the demands of daily life can impose themselves" (CV, 53 years old, Male, CEO).

> "(I have earned the trust of the workers) By showing results, bringing in clients. I think experience also makes the staff respect you and turn to you. I also try to convey to them the history of the company and, above all, how it started" (FB, 47 years old, Male, Sales Manager)

In summary, participants stressed the importance of communicating the company's vision and making employees aware of its history. In order for them to feel an integral part of the company and to commit themselves to the company's objectives.

Learning Based on Mistakes and Experiences

Participants also considered an *approach based on errors and experience*, both to solve challenges and to make decisions, in a way that encourages critical thinking, promotes creativity, autonomy, and self-confidence. Likewise, in this learning process, leaders provide support, active listening, and constant feedback. Additionally, there is corporate training, especially during the first weeks of joining the company, acknowledging the skills and competencies that employees apply.

Also, they give an opening for them to find the solution to problems. "I try to let them be the ones to solve the issues, but I have to admit that I almost always try to be the one who solves the problems" (FBP, 53 years old, Male, Commercial Manager). In case they have difficulties in finding a solution, they are given guidance."I make him reflect so that he can find the solution to the problem on his own. If he doesn't realize it, I guide him in the resolution". (CV, 53 years old, Male, General Manager).

In the same way, there is evidence of learning perceived by staff after being instructed in the subject. "(The staff develops skills) With trainings, I rely on other staff to teach those who don't know what they need to learn." (EM, 55 years old, Female, Manager of Administration and Finance).

> I listen to them, but I let them come to the solution on their own. I always tell them to trust themselves and their abilities, to feel confident, that's the most important thing [...] Of course, you learn from the lessons [...] I try to highlight the lessons learned in each problem and especially how mistakes allow you to improve (JV, 38 years old, Male, Commercial Manager)

In general terms, the development of mental capabilities is encouraged through business training and incorporates the promotion of new approaches, where autonomous learning through error and experience is emphasized.

Exemplary Authority

The leader represents a role model, maintaining the respect and trust of the collaborators through charismatic qualities. In this regard, they indicated that they have an *exemplary authority*, which they promote both for their work experience and professional skills to fulfill their role, for which reason they seek to set a good example with their actions, where they highlight commitment and the importance of mutual respect. “I try to set a good example, I try to make sure that the workers in the company have the same commitment” (FBP, 53 years old, Male, Commercial Manager).

In that sense, interviewees set an example by teaching based on their experience and encouraging values. “(I earn the respect of the staff) Doing the best I can, teaching them to be better every day” (EM, 55 years old, Female, Manager of Administration and Finance). In the same way, they foster values such as respect, respect for others and respect for the environment. “I strive to set a good example, impart positive values, and consistently demonstrate respect” (FB, 47 years old, Male, Sales Manager).

They also report effective customer service and experience in their job functions.“[I earn the respect of the staff] with my ability to attend to customers effectively, to solve problems quickly, and with my overall experience” (CV, 53 years old, Male, General Manager). As can be seen, exemplary authority integrates the influence of the referent in relation to other collaborators, which is why the participants consider maintaining good behavior, promoting values, and demonstrating professional competencies.

Discussion

The purpose of this research is to learn about the experiences of transformational and sustainable leadership from managers of textile companies in Lima. Four relevant thematic patterns were found to explain the leadership experience: individualized leadership, inspirational stimulus, learning based on mistakes and experience and exemplary authority; this partially coincides with what Bass (1985) and Cantón-Mayo et al. (2021) have pointed out regarding sustainable and transformational leadership.

In the individualized leadership, a polite exchange was found between the leader and the worker, where support, empathy, respect, as well as consideration of the collaborator’s work and personal environment prevails. This is related to different research (Erden and Yaşlioğlu 2020; Evitha et al. 2022; Hyun-Duck and Bautista

2022; Saeed and Jun 2021; Ribeiro et al. 2018), who comments that well-being and engagement are promoted if leaders focus on interaction with their collaborators and in areas outside of work. In addition, engagement will optimize the performance and long-term results of companies (Iqbal and Ahmad 2021), which also coincides with what is proposed by sustainable leadership (Cantón-Mayo et al. 2021). On the other hand, the participants pointed out that a leader with a clear supportive behavior is required, a finding that coincides with the study by Hyun-Duck and Bautista (2022). This may be due to the position of manager assumed by the participants, a position that allows them to clearly describe the value of support in transformational and sustainable leadership.

On the other hand, it was found that valuing the employee implies recognizing the abilities of the collaborators and their learning, as well as sharing the company's vision. This is consistent with several studies (Evitha et al. 2022; Ribeiro et al. 2018; Saeed and Jun 2021) which indicate that praising and recognizing work, as well as communicating the organization's vision to employees promotes employee effectiveness and trust. In fact, such appreciation will promote workers' capabilities and improve the social environment, promoting sustainable leadership (Iqbal and Ahmad, 2021; Lämsä et al. 2023). That this was a significant element for the participants may mean that managers emphasize and value results more, which is different in other job positions (Hyun-Duck and Bautista 2022). In this regard, this will contribute to SDG 10 as individualized attention and the recognition of employees' abilities may be related to reducing inequalities (United Nations 2024c).

With respect to inspirational stimulus, managers motivated their employees with an inspiring vision, promoted high expectations, emphasized communication of the vision, and encouraged constant commitment, even outside working hours. This is related to Bass and Avolio (2006), who refer that inspirational encouragement fosters motivation in workers through inspiration and an important emotional bond. Against the aforementioned, Erden and Yaşlioğlu (2020) and Ribeiro et al. (2018) point out that managerial rank personnel can raise expectations and recognition of employees' effort, thus boosting their individual performance through transformational practices. These practices include providing personalized attention, inspiring, promoting intellectual reflection, and providing motivation. This perspective aligns with Leal et al. (2020) and Nsiah-Sarfo et al. (2023), who highlight the importance of skills such as challenging and innovating, managing complexity, thinking long-term, and maintaining an interdisciplinary connection, these being essential capabilities to engage and motivate staff. Additionally, motivating employees through an inspiring vision and high expectations, coupled with a clear company vision, could contribute to the sustainable practices of the company in line with SDG 12 (United Nations 2024d).

In addition, leaders utilize learning from mistakes and experience as a strategy to encourage workers to cultivate critical thinking and creativity for resolving challenges. This is similar to that mentioned by Bass and Avolio (2006) and Mendoza et al. (2014), who refer that a transformational leader highlights the importance of applying resolution skills, which generates critical reflection, as well as continuing professional training in order to ensure better organizational performance. This without failing to identify the strengths of the collaborators, which is assimilated to what is expressed by Erden and Yaşlioğlu (2020), who emphasize the recognition of the worker's situation to implement activities that allow the improvement of their potential, in addition to using feedback.

In this regard, Iqbal and Ahmad (2021) state that a sustainable leader emphasizes the admission of mistakes, and conflict resolution capacity, while other research (Iqbal and Ahmad 2021; Lämsä et al. 2023; Leal et al. 2020; Tang et al. 2022) highlight the strengthening of skills, continuous learning, as well as a vision of long-term results based on sustainability. In this regard, Burawat (2019) refers that when sustainable leaders employ constant training, share ideas, and persuade the will of workers, they will achieve their specific objectives. Thus, companies that implement a combination of technical, operational, and human relations skills and the commitment of employees and leaders will achieve the sustainable development of the company. Given this context, the encouragement of reflection and critical thinking serves as the cornerstone, empowering employees to address challenges and cultivate a proficient workforce, thereby fostering decent working conditions (SDG 8) (United Nations 2024b).

Another of the transcendental aspects for a leader is the influence he or she generates in the personnel, both for his or her professional experience and human capacity through good example, respect and commitment to the company. In this regard, Hyun-Duck and Bautista (2022) state that when leaders meet the needs of male workers (emphasizing equality, honesty, and fairness) and female workers (offering help, providing opportunities for development and rewards), trust, respect, and a positive attitude toward the leader and the organization are generated, which contributes to the sustainability of leadership.

Likewise, Erden and Yaşlioğlu (2020) state that a leader with a work orientation, who encourages participation and decision-making and who proves to be impartial and attentive is appreciated. While Saeed and Jun (2021) point out that leaders must teach by example, so that they alleviate the unequal relationship between them and the collaborators, since, if they maintain a pleasant working environment, the staff will be able to fulfill their duties and integrate into the team, so they must also demonstrate commitment to Corporate Social Responsibility [CSR] so that it is shared (Lämsä et al. 2023). Also, the way that leaders interact with employees contributes to sustainable economic development, aligning with SDG 8 (United Nations 2024b).

Although we have identified significant patterns that shed light on the experiences of transformational and sustainable leadership in textile companies in Lima, it is essential to recognize the limitations of this study. First, difficulties were encountered in accessing participants who met the pre-defined criteria and belonged to companies with similar characteristics. This limitation prevented the possibility of making comparisons based on the participants' positions. Furthermore, information was not collected from other sources or through other instruments that could have allowed methodological triangulation, potentially limiting the depth of the analysis.

Another limitation is linked to the nature of the design used. While the thematic design provides flexibility to identify patterns across the data, a case study design could help provide a detailed analysis of specific companies. This would provide greater depth when analyzing companies rather than focusing exclusively on participants.

Conclusions

Transformational and sustainable leadership in textile companies, from their managers, is manifested through trust and mutual respect of the leader and the collaborator, where skills and support in external aspects of work are recognized. Likewise, the inspiration and communication of the company's vision were elemental to motivate and create a commitment in the workers to the company's objectives. On the other hand, the learning based on mistakes and experience strategy was promoted by means of training, reasoning, creativity, and autonomy. Finally, leadership is assumed as an authority figure who is an example either by his experience and/or personal qualities, which promotes a positive working environment. This will ensure the efficiency and endurance of companies, not just in confronting current obstacles but also in establishing the foundation for long-term productivity and growth. Thus contributing to the Sustainable Development Goals (SDGs), specifically those related to decent work and economic growth (SDG 8), reducing inequalities (SDG 10), and responsible consumption and production (SDG 12).

Moving forward, several recommendations are proposed to enhance the depth of future research. Firstly, considering the challenges encountered in accessing participants and making position-based comparisons, it is recommended to explore alternative research designs that allow for a more diverse participant pool and enable a comparative analysis across different organizational settings. Additionally, to address the limitations related to methodological triangulation, future studies could incorporate information from several sources and utilize different instruments to enrich the analysis.

Moreover, to enhance the robustness of future research, it is recommended to explore qualitative designs, like case study, which delve deeper into the narratives or context of the participants. Furthermore, conducting studies specifically focused on the minor theme identified, such as psychological adaptability, would contribute to a more comprehensive understanding, considering its non-significant impact in the present research.

References

Bass B (1985) Leadership and performance beyond expectations. The Free Press, New York, United States of America

Bass B, Avolio B (2006) Manual for the multifactor leadership questionnaire. Consulting Psychologist Press, Editorial Mindgarden

Bass BM (1998) Transformational and transactional leadership of men and women. In Bass BM (ed) Transformational leadership: industrial, military, and educational impact, pp 72–79. Lawrence Erlbaum Associates

Braun V, Clarke V (2022) Thematic analysis. Sage, A Practical Guide

Burawat P (2019) The relationships among transformational leadership, sustainable leadership, lean manufacturing and sustainability performance in Thai SMEs manufacturing industry. Int J Qual & Reliab Manag. https://doi.org/10.1108/ijqrm-09-2017-0178

Burke RJ, Richardsen AM (2019) Increasing occupational health and safety in workplaces: individual. Edward Elgar Publishing, Work and Organizational Factors

Cantón-Mayo I, García-Martín S, Cañón-Rodríguez R, Grande-de-Prado M (2021) Calidad y Liderazgo Sostenible. Int J Educ LeadShip Manag 9(1):76–91. https://doi.org/10.17583/ijelm.2021.5361

Charoensukmongkol P, Puyod JV (2021) Influence of transformational leadership on role ambiguity and work-life balance of Filipino University employees during COVID-19: does employee involvement matter? Int J Leadersh Educ 27(1):1–20. https://doi.org/10.1080/13603124.2021.1882701

Concejo Directivo Nacional (2017) Código de Ética y Deontología. Colegio de Psicólogos del Perú

Congress of the Republic of Peru (2013) Law that modifies various laws to facilitate investment, promote productive development and business growth, Law 30056. El Peruano. https://cdn.www.gob.pe/uploads/document/file/3017949/Ley%2030056.pdf?v=1649882208

Correia MS (2019) Sustainability: an overview of the triple bottom line and sustainability implementation. Int J Strat Eng 2(1):29–38. https://doi.org/10.4018/IJoSE.2019010103

Creswell J, Poth CN (2018) Qualitative inquiry and research design: choosing among five approaches, 4th edn. SAGE

Daniel BK (2018) Empirical verification of the "TACT" framework for teaching rigour in qualitative research methodology. Qual Res J 18(3):262–275. https://doi.org/10.1108/QRJ-D-17-00012

Del Carpio J (2023) Exportaciones del sector textil y confecciones apuntan a expandirse. Diario El Peruano. https://www.elperuano.pe/noticia/216904-exportaciones-del-sector-textil-y-confecciones-apuntan-a-expandirse

Erden NS, Yaşlioğlu M (2020) Transformational leadership: a qualitative for the practical usage. In Huseyin M, Danis H, Demir E, Ahmet F (eds). Eurasian business perspectives. Proceedings of the 23rd Eurasia business and economics society conference, pp 170 180. Editorial Springer. https://doi.org/10.1007/978-3-030-40160-3

Evitha Y, Sukoco I, Hermanto B, Herawaty T (2022) The importance of transformational leadership in creating work engagement to improve employee performance: a case study of a private university in Jakarta. J Posit Psychol & Wellbeing 6(4):1–14

Gil-Gomez H, Guerola-Navarro V, Oltra-Badenes R, Lozano-Quilis JA (2020) Customer relationship management: digital transformation and sustainable business model innovation. Econ Res-Ekon Istraz 33(1):2733–2750. https://doi.org/10.1080/1331677X.2019.1676283

Hyun-Duck K, Bautista A (2022) Transformational leadership and psychological well-being of service-oriented staff: hybrid data synthesis technique. Int J Environ Res Public Health 19:8981. https://doi.org/10.3390/ijerph19138189

Instituto Nacional de Estadística e Informática (INEI) (2023) Demografía Empresarial en el Perú. I Trimestre del 2023. Instituto Nacional de Estadística e Informática. https://m.inei.gob.pe/media/MenuRecursivo/boletines/boletin_demografia_empresarial_7.pdf

Iqbal Q, Ahmad NH (2021) Sustainable development: the colors of sustainable leadership in learning organization. Sustain Dev 29:108–119. https://doi.org/10.1002/sd.2135

Jauregui-Arroyo RR, Goñi NM, Rondón-Jara E (2022) Estilos de liderazgo de los millennials y su influencia en el desempeño de las pequeñas empresas del sector textil manufactura. Innovar 33(89):147–160. https://doi.org/10.15446/innovar.v33n89.107047

Jiménez A, Villanueva M (2018) Los estilos de liderazgo y su influencia en la organización: Estudio de casos en el Campo de Gibraltar. Gestión Joven 18(1):183–195. https://gestionjoven.org//revista/contenidos_18/13.pdf

Lämsä A, Heikkinen S, Pučėtaitė R (2023) Social sustainability, leadership and human resource management. South Asian J Bus Manag Cases 12(1):7–112. https://doi.org/10.1177/22779779231164518

Leal W, Eustachio JHPP, Caldana ACF, Will M, Lange A, Rampasso IS, Anholon R, Platje J, Kovaleva M (2020) Sustainability leadership in higher education institutions: an overview of challenges. Sustainability 12:3761. https://doi.org/10.3390/su12093761

Mendoza I, García R, Uribe J (2014) Liderazgo y su relación con variables de resultado: un modelo estructural comparativo entre Liderazgo transformacional y transaccional en una empresa de entretenimiento en México. Acta De Investigación Psicológica 4(1):1412–1429. https://doi.org/10.1016/S2007-4719(14)70384-9

Miles M, Huberman M, Saldaña J (2014) Qualitative data analysis: an expanded sourcebook. SAGE

Muralidharan E, Pathak S (2018) Sustainability, transformational leadership, and social entrepreneurship. Sustainability 10(2):567. https://doi.org/10.3390/su10020567

Nosratabadi S, Mosavi A, Shamshirband S, Kazimieras Zavadskas E, Rakotonirainy A, Chau KW (2019) Sustainable business models: a review. Sustainability 11(6):1663. https://doi.org/10.3390/su11061663

Nsiah-Sarfo DJ, Ofori D, Agyapong D (2023) Sustainable procurement implementation among public sector organizations in Ghana: the role of institutional isomorphism and sustainable leadership. Clean Logist Supply Chain 8:100118. https://doi.org/10.1016/j.clscn.2023.100118

Palafox-Soto MO, Ochoa-Jimenez S, Jacobo Hernandez CA (2021) El liderazgo y su relación con la cultura organizacional en las pequeñas y medianas empresas de la ciudad de Obregon, Sonora. Nova Scientia 13(26):00017. https://doi.org/10.21640/ns.v13i26.2682

Prasad KDV, Vaidya RW, Mangipudi MR (2020) Effect of occupational stress and remote working on psychological well-being of employees: an empirical analysis during covid-19 pandemic concerning information technology industry in hyderabad. Indian J Commer Manag Stud 11(2):01–13. http://doi.org/10.18843/ijcms/v11i2/01

Purwanto A (2021) The effect of transformational leadership, job satisfaction and organizational commitments toward organizational citizenship behavior of Indonesian senior high school teachers. Ilk Online-ElemTary Educ Online 20(3):177–185. https://doi.org/10.17051/ilkonline.2021.03.17

Ribeiro N, Yücel Í, Roque D (2018) How transformational leadership predicts employees' affective commitment and performance. Int J Product Perform Manag 67(9):1901–1917. https://doi.org/10.1108/IJPPM-09-2017-0229

Saeed F, Jun Y (2021) The influence of transformational leadership on employee retention in Pakistan: function of job embeddedness as mediator- international journal of management. Account Econ 8(12):867–881. https://doi.org/10.5281/zenodo.6390898

Swain J (2018) A hybrid approach to thematic analysis in qualitative research: using a practical example. SAGE Res Methods Cases. https://doi.org/10.4135/9781526435477

Tang Y, Chen Y-J, Shao Y-F, Cao Q (2022) The impact of sustainable transformational leadership on sustainable innovation ambidexterity: empirical evidence from green building industries of China. Front Public Health 10. https://doi.org/10.3389/fpubh.2022.814690

Tello C (2019) Sector textil y confecciones exportó US$ 1,400 millones. Diario El Peruano. https://elperuano.pe/noticia/76728-sector-textil-y-confecciones-exporto-us-1400-mllns

United Nations (2024a) The sustainable development agenda. https://www.un.org/sustainabledevelopment/development-agenda/

United Nations (2024b) Sustainable development goals. Goal 8: promote inclusive and sustainable economic growth, employment and decent work for all. https://www.un.org/sustainabledevelopment/economic-growth/

United Nations (2024c) Sustainable development goals. Goal 10: reduce inequality within and among countries. https://www.un.org/sustainabledevelopment/inequality/

United Nations (2024d) Sustainable development goals. Goal 12: ensure sustainable consumption and production patterns. https://www.un.org/sustainabledevelopment/sustainable-consumption-production

Van Diggele C, Burgess A, Roberts C, Mellis C (2020) Leadership in healthcare education. BMC Med Educ 20(2):456. https://doi.org/10.1186/s12909-020-02288-x
World Bank (2023a) The World Bank in Latin America and the Caribbean. The World Bank. https://www.worldbank.org/en/region/lac/overview
World Bank (2023b) The World Bank in Peru. The World Bank. https://www.worldbank.org/en/country/peru/overview
World Bank (2022) World development report 2022: finance for an equitable recovery. World Bank. https://doi.org/10.1596/978-1-4648-1730-4

Chapter 5
Obstacles to Innovation in Textile SMEs in the Usaquen District of Bogota, Colombia

Juan Carlos Olaya Molano, Angie Lorena Romero Quiroga, Carolina Aguirre Garzón, and Ernesto León-Castro

Introduction

Taking into account that since 2020 different world events have occurred, among others the sanitary emergency caused by Covid-19, the war between Russia and Ukraine and the war between Israel and Gaza brought adverse effects to the world economy, such as currency devaluation, unemployment, bankruptcy of companies especially smaller ones and therefore less capital, so much so that "organizations have needed to innovate to remain in the market, which is increasingly competitive, dynamic and challenging" (Isaza et al. 2022), however, it should be clarified that "micro, small and medium enterprises have unique characteristics that allow them to foster technological innovations in diverse economic environments, which may include improvements in organizational structure, more fluid communication both internally and externally, and greater control in decision-making" what makes relevant the study on the perception of the obstacles they have to innovate (Ramos-Rivadeneira and Jiménez-Toledo 2024).

Innovation is a crucial aspect of companies nowadays, and it can be defined as "the introduction of something new (idea, method, device, product, service, among

J. C. O. Molano · A. L. R. Quiroga
Universidad de San Buenaventura, Bogotá, Colombia
e-mail: jolaya@usbbog.edu.co

A. L. R. Quiroga
e-mail: inv.profesional@usbbog.edu.co

C. A. Garzón (✉) · E. León-Castro
Universidad Católica de La Santísima Concepción, Concepción, Chile
e-mail: caguirre@doctorado.ucsc.cl

E. León-Castro
e-mail: eleon@ucsc.cl

A. E. Toscano-Hernández et al. (eds.), *Sustainability, Stakeholders and Marketing in the Textile Sector*, SDGs and Textiles, https://doi.org/10.1007/978-981-97-6952-0_5

others)" (Kahn 2018; Rogers and Rogers 1998). Also, the Royal Spanish Academy (2024) defines it as "the creation or possible modification of a product, made for commercial purposes or for introducing it to the market". Even when innovation has been designated as one of the most critical factors for a company to succeed (Am et al. 2020; Kuratko et al. 2014), many obstacles make it difficult for this to happen (Blanchard et al. 2013; de-Oliveira and Rodil-Marzábal 2019) being financial constraints one of the main obstacles (Błach 2020; Godke Veiga and McCahery 2019), companies need to understand and know the main obstacles that will affect the innovation process (Akhmadi and Tsakalerou 2020), this will allow them to improve the way in which processes are designed and find a way to address the barriers that exist within the industry.

In turn, it should be noted that in the Colombian context, concerning innovation, the country ranked third in the Global Innovation Index among the 18 economies of Latin America and the Caribbean (Startup Genome 2024). However, the World Intellectual Property Organization (WIPO) (2021) identifies that it still has a relatively unbalanced innovation system, among other factors of lower performance in human capital development and research, in contrast to a reasonably good performance in market sophistication and business sophistication.

Given the importance of Colombian SMEs, Bancoldex (2024) defines them as a "group of small and medium-sized companies with total assets of more than 500 and up to 30,000 Minimum Legal Wages in Force". In addition, in Colombia, microenterprises represent 95.3%, small enterprises 3.5%, and medium and large enterprises account for 0.9% and 0.3% of the national total, respectively. This demonstrates the importance of analyzing small, micro, and medium-sized (SME) companies since they are essential in the country's economic development.

One of the most important industrial sectors in the world is textile, which was the first to embrace the Industrial Revolution, and its impact on the economies of different countries depends on its evolution and innovation (Puig et al. 2022; Shishoo 2012). The size of the global textile market was valued at USD 1,695,130 million in 2022 (Grand View Research 2023), and specifically in the case of Colombia, the textile industry represents 9.4% of the Gross Domestic Product (GDP) of the country's manufacturing sector, employs more than half a million people and generates more than COP 20 billion in annual revenue (Econexia 2023).

Innovation processes cannot be unlinked from sustainability, defined as "development that is capable of meeting the needs of present generations without compromising the ability of future generations to meet their own needs" (Eurofins 2023). The textile sector is closely related to the Social Development Goals (SDG) defined by the United Nations, especially SDG 8, "Decent work and economic growth" target 8.4, which is defined as: "Improve progressively, through 2030, global resource efficiency in consumption and production and endeavor to decouple economic growth from environmental degradation, in accordance with the 10-Year Framework of Programmes on Sustainable Consumption and Production, with developed countries taking the lead" (United Nations 2024a) because when there is growth in the industry under analysis, there will be an increase in the number of jobs, with better economic conditions and in itself there will be an increase in the economy of the company, which will result in the economic and social growth of the country, but always taking care of the environmental degradation and changing the consumer behavior about fast fashion or just one season clothes.

Another related SDG that must be taken into account in the innovation processes in the textile sector is number 9, "Industry, innovation and infrastructure", specifically target 9.4 which is: "By 2030, upgrade infrastructure and retrofit industries to make them sustainable, with increased resource-use efficiency and greater adoption of clean and environmentally sound technologies and industrial processes, with all countries taking action in accordance with their respective capabilities" (United Nations 2024b). This is achieved in the textile industry if innovative processes are carried out to make organizations' processes more efficient while caring for the resources used to make the different products. Finally, SDG 12, "Responsible production and consumption," with its target 12.a: "Support developing countries to strengthen their scientific and technological capacity to move towards more sustainable patterns of consumption and production" is highly related to the textile sector in Colombia (United Nations 2024c). This goal is considered because customers are expected to take positions of environmental care that will be reflected in the sustainability of the sector, the country, and the planet.

Due to the economic, social, and environmental impact generated by the textile sector in Colombia, it is important to know the main obstacles that this type of company must overcome in their life in the market. Based on the above, the study's central question is: What are the obstacles to innovation in SMEs in the textile sector in the Usaquén district of Bogotá, Colombia?

The paper is organized as follows. Section 5.2 presents the theoretical framework of innovation and sustainability; Sect. 5.3 presents the detailed methodology used in this work; Sect. 5.4 presents the results and discusses the main findings. Finally, Sect. 5.5 summarizes the main conclusions of the paper.

Theoretical Framework

Schumpeter (1934) a pioneer in the study of innovation, defines it as the establishment of a new production function. This definition is guided by the principle of creative destruction, as the ability of organizations to develop incentives to introduce innovations. On the other hand, and as a complement, the Organization for Economic Co-operation and Development (OECD) (2023) has defined innovation as: "a new or enhanced product or business process of an enterprise that produces favorable environmental outcomes or reduces adverse impacts when contrasted with the firms earlier product or business process, and that is accessible to potential users or has been put into practice". In line, León (2023) indicates that companies' innovation performance is based on three dimensions: development of incremental innovation, development of radical innovation, and opening of new markets. The first consists of improving the organization's goods and/or services, the second consists of producing new goods and services, and the third consists of the entrepreneurial capacity to reach new markets.

On the other hand, the Oslo Manual published by OECD & Eurostat (2018) establishes that: "An innovation barrier prevents a non-innovative firm from engaging in innovation activities or an innovation active firm from introducing specific types of innovation. Innovation obstacles increase the cost or create technical problems". One example of an innovation obstacle is identified by Méndez-Morales & Yanes-Guerra (2018) highlighting that investors can only appropriate the private benefits of their projects, so they will only be willing to finance those that demonstrate favorable rates of return. This complicates the financing of research, development, and innovation (R&D&I) projects due to the high rate of uncertainty that exists, in addition to the need for financing that Colombian SMEs have (Sierra-González et al. 2021).

For this reason, the public and private sectors seek to identify the obstacles that prevent significant progress in innovation to establish strategies and policies that generate progress in the country, particularly in the business sector. Thus, the National Administrative Department of Statistics (DANE for its acronymous in Spanish) (2021) in the Technical Bulletins: Survey of Technological Development and Innovation in the Manufacturing Industry (EDIT) 2019–2020 identifies some obstacles such as difficulty in accessing external financing, scarcity of own resources, scarce information on public support instruments, uncertainty about the demand for innovative goods and services, uncertainty about the success of the technical implementation of the project and the low profitability of innovation.

Another issue affecting all businesses in general is how they approach sustainability in their business models, and the textile industry is no exception. Authors such as Alkaya and Demirer (2014) have reviewed the environmental practices and technologies used in the industry, Hasanbeigi and Price (2012) presented 184 energy efficiency measures for the sector, and Chen et al. (2021) analyze the way clothes end up in landfills at a high speed and present some strategies to reduce that such as renewable raw materials, rethinking production, maximum use and reuse of textile products and recycling strategies.

As can be seen, different authors have analyzed how the textile industry can improve their practices and, because of that, is highly related to the SDGs, such as the research of Akter et al. (2022) that presents a model to achieve SDG 12 for textile manufacturing and material waste management in Bangladesh. Also, Papú Carrone (2020) indicates that to make a significant change to achieve SDG 12 is to recognize traceability and transparency in the textile industry patterns.

In the case of SDG 8, Scherrer (2020) indicates that changes in the different policies must be made to reach sustainable development because the oversupply of the working-age population limits, industrial policies, and infrastructural bottlenecks have become a limitation to improve the working conditions of the textile industry. Finally, Alam et al. (2023) present a Pareto analysis of the barriers to achieving SDG 9 in the textile sector; in their results, they found 19 factors, being the most important the lack of government support policies, insufficient incentives & heavy taxation and inadequate capital investment.

With all this information, innovation is needed to have a better textile industry, not only to improve the financial results but also to be related to the SDG to increase the real value of the textile companies. Because of that, an analysis of the different obstacles that the industry faces is relevant to improve how they visualize their limitations and strengths, and with that information, they can start to make new processes and policies for the sector.

Methodology

The research was conducted through the quantitative methodology, which allowed the collection of pertinent information to interpret it later and thus be able to present the analysis of the results obtained to suggest some strategies that contribute to the increase of innovation practices in the SMEs of the textile sector in the locality of Usaquén initially and in the medium term the research can be replicated to other sectors, both industrial and geographic, which allows increasing the levels of innovation in the textile industry in Colombia. The research is of a descriptive type, whose purpose is to present the characteristics of the sample found at the time of the application of the measurement instrument. The type of cut used in the research was transversal, meaning the instrument was applied only once in the defined sample. The instrument used for the study was a closed form with multiple selection options. The survey was applied to 72 SMEs out of 2720 in the textile sector in the locality of Usaquén, with a confidence level of 90% (DANE 2021).

To obtain the information, a questionnaire validated by three academic experts and three microentrepreneurs of the corresponding sector was used, which was structured on a Likert scale with five response options that made it possible to calibrate the different degrees of importance, ranging from minor to major, these being: not very important, slightly important, important, very important and extremely important. Although the survey was structured with twenty-four closed questions, only nine were analyzed for this article due to the degree of relevance.

Results

Considering the methodology and the article's purpose, the results were structured according to the type of obstacle presented in the textile MSMEs of Usaquén. Hence, Table 5.1 shows the results of the financial area. According to these, it can be seen that the financial variable is seen as an obstacle that is too important for 37.5% and very important for 27.8% of the MSMEs in the sector. This suggests that 65.3% of the respondents give relevance to the variable since most companies observe that the financial challenges that should be given the highest priority in the organization are the availability of funds as working capital and return on investment, which leaves funds to generate organizational innovation in second place, especially because of

Table 5.1 Obstacles related to financing

	N	%
Unimportant	7	9.7
Less important	7	9.7
Important	11	15.3
Very important	20	27.8
Too important	27	37.5

Source Own elaboration (2024)

the high risk of uncertainty in return on investment. In addition, a high percentage of SMEs are financed through the financial sector, which forces them to cover somewhat high-interest rates and, in most cases, in the short term. At the same time, the expected results of innovation in an optimistic scenario occur in a high percentage in the medium and/or long term.

For the obstacle related to financing, the results show an arithmetic mean of 14.4 and a standard deviation of 8.82, results that show that for 19.14% of the textile entrepreneurs this variable is not considered important, because although the level of competition is high, the levels of innovation are not so high, while for 65.1% of the entrepreneurs it does have a high level of relevance, because financial resources are considered a limitation to enter into high-trend innovation processes and knowledge.

Another of the obstacles analyzed in the research was knowledge. Table 5.2 presents the analysis considering the knowledge the collaborators must have to carry out innovative activities in the SMEs where they work. SMEs in the textile sector in the locality analyzed perceive knowledge limitations as a relevant obstacle to innovation development (71% of the answers were very or too important). This shows that companies with unskilled personnel should invest in training in their area of specialization and innovation concepts so that, by increasing knowledge and skills, elements of improvement can be presented in the organizations.

The results obtained in Table 5.2 show an arithmetic mean of 14.4 and a standard deviation of 8.414, which shows that there is a high degree of variability, this is due to the fact that on the one hand 9.7% of the sample investigated does not consider knowledge limitations important, because many of the MSMEs continue working in a traditional way, licensing free technology and on the other hand 71% give it a

Table 5.2 Obstacles related to knowledge limitations

	N	%
Unimportant	10	N/A
Less important	6	9.7
Important	12	19.4
Very important	28	45.2
Too important	16	25.8

Source Own elaboration (2024)

Table 5.3 Obstacles related to knowledge barriers

	N	%
Unimportant	15	N/A
Less important	5	6.9
Important	11	15.3
Very important	29	40.3
Too important	12	16.7

Source Own elaboration (2024)

high level of importance, because they feel the impact of competitors that do carry out structured innovation activities and even more so when they have international companies as direct competitors.

Table 5.3 presents the surveyed entrepreneurs' perception of the barriers their establishments encounter in receiving specialized knowledge through private and public organizations. Knowledge barriers are perceived by the sector's entrepreneurs as a notable obstacle since the answers given by 57% of interviewees show concern by considering it very and too important. This suggests that public and private entities should prioritize access to updated information and training within the sector's needs. It should be noted that there are different training options, especially in private higher education institutions, through extension and postgraduate courses. Even so, these generate a significant demand for money that many SMEs cannot afford to subsidize. Innovation parks are another training model to which a collaborator or micro-entrepreneur can be linked; however, the drawback arises from the number of places, given the high level of demand. The same is true of public institutions of higher education. Although they offer some courses on the subject, the number of places to take them is quite small compared to the existing demand.

The data collected is in Table 5.3. They show an arithmetic mean of 14.4 and a standard deviation of 8.9, which allows establishing a high level of variability, this is due to the fact that only 6.90% of the entrepreneurs indicated that knowledge barriers are not important, this is due to the low level of innovation existing in the country because traditional models that have been working in the organizations are still being adopted. On the contrary, 57% of the respondents stated that knowledge barriers are very important for the development of innovative activities in the MSMEs of the textile sector in the town of Usaquén, because they consider the existence of some limits that do not allow improving knowledge regarding innovation processes and therefore cannot be implemented in the MSMEs under analysis.

Another obstacle reviewed in the research was the regulations that seek to increase the levels of innovation in Colombian companies, in this case, SMEs. Table 5.4 presents the results obtained. Regulations are perceived as a moderate obstacle to the development of innovation processes by 46% of the surveyed sample, which evidences the existence of challenges and opportunities that may arise from regulations and legal requirements that may limit or complicate the implementation of

Table 5.4 Obstacles related to regulations

	N	%
Unimportant	15	N/A
Less important	8	11.1
Important	16	22.2
Very important	20	27.8
Too important	13	18.1

Source Own elaboration (2024)

innovation. If handled effectively and efficiently, they can become opportunities for organizations that process them to their advantage.

Table 5.4 shows the results obtained in the application of the survey, showing an arithmetic mean of 14.4 and a standard deviation of 4.39, which allows us to establish a high level of variability. This is due to the fact that 11.10% consider regulations as an unimportant factor for obtaining innovative processes at advanced levels. This is due to the fact that the entrepreneurs comply with the regulations that affect them directly, such as the payment of tax obligations, incorporation of the organization, and labor obligations, in addition to the fact that in Colombia there is not a wide range of regulations regarding innovation processes. On the other hand, 45.9% consider the existence of and compliance with the regulations governing innovative business processes to be very important, since it will allow them to increase their level of competitiveness in the local, national, and international markets.

The last obstacle to be analyzed is the country's lack of technological level, which is perceived as a moderate obstacle to innovation (46% of responses are very or too important). This suggests that the level of technological development and infrastructure available in the country can influence a company's ability to innovate. This obstacle could be particularly relevant for companies that rely heavily on advanced technology for their innovation.

The arithmetic mean is 14.4 and the standard deviation is 5.31, which determines a high level of variability in the answers of the respondents, this is due to the fact that only 8.30% of the population considers the technological level of the MSMEs to be of little importance for the development of their innovative activities, this is given in Colombia taking into account the artisanal and/or manufacturing production model that has been in place throughout history. On the contrary, 55.9% consider the technological level of the MSMEs to be very important because the higher the technological level, the more precise innovation activities can be carried out, which is reflected in the level of competitiveness of the MSMEs in the textile sector.

As seen with all the results presented in Table 5.5, the ranking of obstacles based on the information provided by the Usaquén SMEs is knowledge limitations, finances, knowledge barriers, regulations, and technological level. An interesting result is that knowledge limitations, not the financial aspect, are the main perceived obstacle. Entrepreneurs are highly concerned about receiving adequate training for their employees as they consider that this limits innovation, which, if interpreted

Table 5.5 Obstacles related to technological level

	N	%
Unimportant	17	N/A
Less important	6	8.3
Important	16	22.2
Very important	20	27.8
Too much important	13	18.1

Source Own elaboration (2024)

together with the 57% importance of knowledge barriers, an evident interest of the sector to improve their training processes can be discerned. In turn, this can be linked to the possibility of governments generating public policies related to the adequate training of organizations, which would allow them to improve companies' innovation levels.

This situation is related to the 46% result of deep concern of companies about regulations, as this shows that SMEs still have a latent concern about having regulations that allow them to improve innovation rates within their organizations. Last but not least, 46% are concerned about technological obstacles, which, given the new realities generated after COVID-19, means that SMEs need to join the digital world increasingly.

An important aspect to consider is that this research has certain limitations that should be detailed. The first relates to the fact that the results are specific to Usaquén in Bogotá, Colombia, within the textile sector, so the results cannot be generalized to all the companies in the textile sector or the area studied. In turn, these results respond to a sample survey with a 90% confidence level. However, we consider that these limitations do not affect the quality of the study, whose objective was to study the reality of the textile companies in the area and, given this, to know the reality of the sector specifically and not a generalized reality that does not respond to what happens at the local or regional level.

On the other hand, it is worth exploring the innovative processes developed by the company in the field of digital marketing. Before delving into the analysis of these processes in the context of the SMEs studied, it is essential to highlight that social networks are of considerable importance for 45.8% of the sample surveyed as a valuable resource in digital marketing. This is attributed to their effectiveness in reaching a wide audience and fostering direct customer engagement.

Interestingly, only 26.4% of companies have a website, a shallow figure considering that a website is a critical element in establishing an online presence. Websites play a key role in e-commerce, lead generation, and branding. However, it must be recognized that maintaining a website requires a substantial investment of time and financial resources.

In addition, it is crucial to examine the limited adoption of certain digital marketing strategies. Internet stores, search engine optimization (SEO), and email marketing are used by a small minority of companies (5.6%, 9.7%, and 5.6%, respectively).

This discrepancy could indicate the need for specialized knowledge or resources many companies lack. Another possibility is that these strategies are not effectively tailored to specific business models or audiences.

On a related note, a slightly higher percentage of companies (11.1% and 8.3%, respectively) incorporate search engine marketing (SEM) and affiliate marketing into their approach. However, these percentages remain in the minority. The specialized skill set and the significant investment required for these methods could explain their low uptake.

In addition, a relatively small number of companies are adopting content marketing, using digital directories, and maintaining blogs (8.3%, 23.6%, and 8.3%, respectively). While these digital marketing strategies can produce remarkable results in terms of traffic generation, SEO improvement, and brand reputation, they require a significant investment of time and expertise at the same time.

Most companies have not yet integrated most of these innovative processes. For example, 90.3% of companies have yet to incorporate more advanced customer analytics. This observation may signify a shortage of resources or expertise needed to conduct in-depth data analytics, a crucial component in gaining better customer insight and personalizing the overall customer experience (see Table 5.6).

Also, 87.5% of companies have not yet adopted personalized and segmented campaigns. This strategic approach is critical in digital marketing, as it significantly improves the effectiveness of advertising campaigns and elevates the overall customer experience (see Table 5.7).

79.2% of companies have not implemented better processes and multiple payment methods. This suggests that many companies may be missing opportunities to simplify the payment process and offer customers more payment options, which can impact customer satisfaction and sales (see Table 5.8).

The use of virtual reality in digital marketing is even scarcer, with a staggering 95.8% of companies yet to integrate this technology. Given that virtual reality is still an evolving innovation and often requires a substantial financial commitment, it is unsurprising that its adoption is not widespread among many companies (see

Table 5.6 More complete customer analysis

	N	%
0	65	90.3
1	7	9.7

Source Own elaboration (2024)

Table 5.7 Personalized and segmented campaigns

	N	%
0	63	87.5
1	9	12.5

Source Own elaboration (2024)

Table 5.8 Improved processes and various payment methods

	N	%
0	57	79.2
1	15	20.8

Source Own elaboration (2024)

Table 5.9). Taken together, these results show that there is considerable scope for companies to adopt and benefit from innovative digital marketing practices.

Finally, it is important to indicate that this study has some limitations; the main one is that the results cannot be generalized, and they represent the reality of the textile SME companies in Usaquen, Colombia. The other limitation of the study is that the results can change if the companies in the survey change, even when the sample is a statistically significant representation of the population; changing the companies or the people who respond to the survey may change. Because of that, it will be interesting to conduct the same survey on different textile clusters in different countries. This will allow us to analyze the realities of the obstacles in different regions.

Considering the information presented in Tables 5.1, 5.2, 5.3, 5.4 and 5.5, Fig. 5.1 presents the information in a comparative way to establish the existing relationship or not between the different variables under analysis. The information shows that the obstacles considered most important by the sector's microentrepreneurs for the development of their activities are limited knowledge and knowledge barriers, because the acquisition of knowledge on innovation issues is not open to small entrepreneurs due to the high costs involved in obtaining it. Open access knowledge is not updated in most cases and in some others, it is administered by international institutions or national governments, which makes it very difficult to acquire.

The training processes, as indicated above, are carried out on the trends that are currently in the country, which are clearly quite old with respect to international economies and are quite costly for the micro-entrepreneurs in the sector under analysis. On the other hand, the large entrepreneurs, having greater economic power, can obtain more advanced knowledge of the innovation processes that are being structured at the international level, which means that there is a large gap in relation to the micro-entrepreneurs in the sector. At the same time, for microentrepreneurs in the sector, the use of the English language becomes another major drawback, since a high rate of information is in this language, which makes it more difficult to acquire knowledge through the review of literature or participation in an international event.

Table 5.9 Virtual reality

	N	%
0	69	95.8
1	3	4.2

Source Own elaboration (2024)

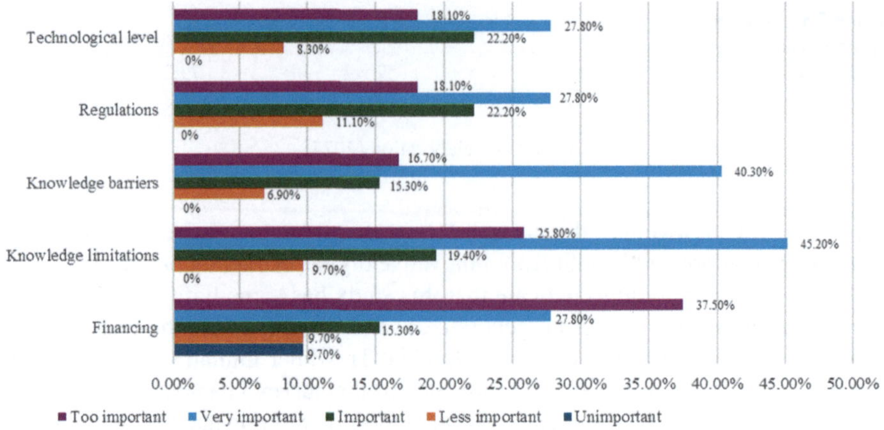

Fig. 5.1 Comparative. *Source* Own elaboration (2024)

In addition, entrepreneurs indicate that another major obstacle is financing, due to the fact that, as indicated above, the SMEs observed have small capital in relation to national and international competition, which means that they cannot finance state-of-the-art training processes, in addition to not being able to acquire machinery, infrastructure and, in general, assets of a high technological level due to the prices they have. In addition, the sources of financing are quite expensive, therefore it is complicated to finance through the national financial system, or if the payment of installments is made, it does not allow the growth of the MSMEs that take the credit. On the other hand, the labor force does not have high qualification standards, firstly due to the payment of salaries made by the entrepreneurs and secondly because the processes of acquiring knowledge on innovation are quite expensive and in many occasions, the collaborator is the one who has to finance them in order to remain competitive in the environment, while on other occasions the training is provided by the entrepreneurs, but as indicated above they are not the most updated compared to other national or foreign organizations.

Conclusions

The central question of this research was: What are the obstacles to innovation in SMEs in the textile sector in the Usaquén district of Bogotá, Colombia? This question was answered through a survey of 72 SMEs in that district. Among the main findings, it became evident that the main obstacle to innovation in the textile sector is knowledge limitations, even more so than financial issues. This is an interesting finding, both for managers and government, since they can focus on generating public policies to solve this situation and thus increase the levels of innovation in the organization.

On the other hand, it was detected that regulatory issues are perceived as a major obstacle, so if this is analyzed together with the concern for knowledge, a relevant area of opportunity can be identified for universities, which can support training for the sectors and this would be a process that would allow to address a social problem more comprehensively. At the same time, it was found that there is an important opportunity within Digital Marketing since there is not much knowledge about the correct use of tools such as social networks, which are used to upload information but not as a process of obtaining and analyzing data.

With all the results obtained, is clear that the textile industry needs to improve in their internal processes and the government must make changes in the policies that are being applied nowadays. As has been done in different SDG studies for the sector, in Usaquen, Colombia, some important challenges remain to achieve different goals such as the 8, 9, and 12. An example is how the companies have a knowledge problem because they have been doing the same processes for years, without considering the new technology, industrial trends, labor conditions, and environmental protection. If these conditions are not included in the latest strategies the companies are doing, achieving the SDG goals won't be easy.

Finally, there are relevant future lines of research, such as data analysis, to generate predictive models to visualize which strategies would be the most effective for the textile sector and which would use fewer resources with more efficient results (Espinoza-Audelo et al. 2020; Fonseca-Cifuentes et al. 2021; Ruiz-Morales et al. 2021). On the other hand, it is considered that the textile sector is of vital importance not only for Colombia but also for the world economy, so an analysis of ESG (environmental, social, and governance) processes within this sector, which has been classified as highly polluting, so research should be focused on supporting it to achieve adequate solutions (Akbar 2014; Yousaf et al. 2021).

Acknowledgements Research supported by Red Sistemas Inteligentes y Expertos Modelos Computacionales Iberoamericanos (SIEMCI), project number 522RT0130 in Programa Iberoamericano de Ciencia y Tecnologia para el Desarrollo (CYTED).

References

Akbar A (2014) Corporate governance and firm performance: evidence from textile sector of Pakistan. J Asian Bus Strat 4(12):200–207

Akhmadi S, Tsakalerou M (2020) Obstacles to innovation–is there a need for consensus? pp 1–6

Akter MMK, Haq UN, Islam MM, Uddin MA (2022) Textile-apparel manufacturing and material waste management in the circular economy: a conceptual model to achieve sustainable development goal (SDG) 12 for Bangladesh. Clean Environ Syst 4:100070

Alam MFB, Hosen MI, Mridha JH, Chowdhury SE, Rahman MA (2023) Assessing the barriers of integrating technological innovations in textiles sector: implications towards sustainable production. Green Technol Sustain 1(3):100039

Alkaya E, Demirer GN (2014) Sustainable textile production: a case study from a woven fabric manufacturing mill in Turkey. J Clean Prod 65:595–603

Am JB, Furstenthal L, Jorge F, Roth E (2020) Innovation in a crisis: why it is more critical than ever. McKinsey & Company
Błach J (2020) Barriers to financial innovation—corporate finance perspective. J Risk Financ Manag 13(11):273
Blanchard P, Huiban J-P, Musolesi A, Sevestre P (2013) Where there is a will, there is a way? Assessing the impact of obstacles to innovation. Ind Corp Chang 22(3):679–710
Chen X, Memon HA, Wang Y, Marriam I, Tebyetekerwa M (2021) Circular economy and sustainability of the clothing and textile industry. Mater Circ Econ 3:1–9
DANE (2021) Encuesta de Desarrollo e Innovación Tecnológica Industria manufacturera (EDIT X). https://www.dane.gov.co/files/investigaciones/boletines/edit/boletin_EDIT_manufacturera_2019_2020.pdf
de-Oliveira F, Rodil-Marzábal Ó (2019) Structural characteristics and organizational determinants as obstacles to innovation in small developing countries. Technol Forecast Soc Change 140:306–314
Econexia (2023) Conoce las principales tendencias del sector textil para el 2023. https://econexia.com/es/contenidos-articulo/moda/1223/Conoce-principales-tendencias-sector-textil-para-el-2023
Espinoza-Audelo LF, Olazabal-Lugo M, Blanco-Mesa F, León-Castro E, Alfaro-Garcia V (2020) Bonferroni probabilistic ordered weighted averaging operators applied to agricultural commodities' price analysis. Mathematics 8(8):1350
Eurofins (2023) ¿Qué es la Sostenibilidad y el desarrollo sostenible? Eurofins Environment Testing Spain. https://www.eurofins-environment.es/es/sostenibilidad-desarrollo-sostenible/
Fonseca-Cifuentes G, León-Castro E, Blanco-Mesa F (2021) Predicting the future price of a commodity using the OWMA operator: an approximation of the interest rate and inflation in the brown pastusa potato price. J Intell & Fuzzy Syst 40(2):1970–1981
Startup Genome (2024) The global startup ecosystems report 2023. Startup Genome. https://startupgenome.com/report/gser2023
Godke Veiga M, McCahery JA (2019) The financing of small and medium-sized enterprises: an analysis of the financing gap in Brazil. Eur Bus Organ Law Rev 20:633–664
Grand View Research (2023) Textile market size, share & trends analysis report by raw material (wool, chemical, silk), by product (natural fibers, polyester), by application, by region, and segment forecasts, 2023—2030. https://www.grandviewresearch.com/industry-analysis/textile-market
Hasanbeigi A, Price L (2012) A review of energy use and energy efficiency technologies for the textile industry. Renew Sustain Energy Rev 16(6):3648–3665
Isaza JF, Kruger SD, Zanin A, Márquez VJB, Coa LMD, Zanella C, Conte A (2022) Caracterización en la gestión de innovación de pymes por efecto del Coronavirus: Estudio comparativo Colombia y Brasil. Revista Estrategia Organizacional 11(1):87–109
Kahn KB (2018) Understanding innovation. Bus Horiz 61(3):453–460
Kuratko DF, Covin JG, Hornsby JS (2014) Why implementing corporate innovation is so difficult. Bus Horiz 57(5):647–655
Leon OA (2023) Impacto de las capacidades de big data en la innovación empresarial. Ingenieria y Competitividad 25(2)
Méndez-Morales A, Yanes-Guerra C (2018) Financial systems and private innovation activity. A research for OECD countries. In: RAIS conference proceedings-the 11th international RAIS conference on social sciences
OECD (2023) OECD business innovation indicators 2023 edition. https://www.oecd.org/sti/inno/innovation-indicators-2023-highlights.pdf. https://www.oecd.org/sti/inno/innovation-indicators-2023-highlights.pdf
OECD & Eurostat (2018) Oslo manual 2018. https://www.oecd-ilibrary.org/content/publication/9789264304604-en

Papú Carrone N (2020) Traceability and transparency: a way forward for SDG 12 in the textile and clothing industry. The UN sustainable development goals for the textile and fashion industry, pp 1–19
Puig F, Cantarero S, Verdone F (2022) Coronavirus versus the textile industry: cluster lessons for future challenges. Fash Text 9(1):10
¿Qué es una pyme? (2024). https://www.bancoldex.com/es/que-es-una-pyme-1338
Ramos-Rivadeneira DX, Jiménez-Toledo JA (2024) La innovación desde las tecnologías emergentes para la competitividad empresarial. Gestión y Desarrollo Libre 9(17)
Rogers M, Rogers M (1998) The definition and measurement of innovation, vol 98. Melbourne Institute of Applied Economic and Social Research Parkville, VIC
Royal Spanish Academy (2024) Innovacion. "Diccionario de la lengua española"-Edición del Tricentenario. https://dle.rae.es/innovaci%C3%B3n
Ruiz-Morales B, Espitia-Moreno IC, Alfaro-Garcia VG, Leon-Castro E (2021) Sustainable development goals analysis with ordered weighted average operators. Sustainability 13(9):5240
Scherrer C (2020) Superfluous workers: why SDG 8 will remain elusive, pp 119–135
Schumpeter JA (1934) The theory of economic development: an inquiry into profits, capital, credit, interest, and the business cycle
Shishoo R (2012) The importance of innovation-driven textile research and development in the textile industry. In: The global textile and clothing industry, pp 55–76. Elsevier
Sierra-González JH, Londoño-Bedoya DA, García-Ospina JM (2021) Innovation financing in Colombia: an explicative proposal. Cuadernos De Administración 34:1–21
United Nations (2024a) Goal 8. Department of Economic and Social Affairs. https://sdgs.un.org/goals/goal8#targets_and_indicators
United Nations (2024b) Goal 9. Department of Economic and Social Affairs. https://sdgs.un.org/goals/goal9#targets_and_indicators
United Nations (2024c) Goal 12. Department of Economic and Social Affairs. https://sdgs.un.org/goals/goal12#targets_and_indicators
WIPO (2021) Global innovation index 2021: tracking innovation through the COVID-19 crisis. https://www.wipo.int/edocs/pubdocs/en/wipo_pub_gii_2021.pdf
Yousaf U, Ali SA, Ahmed M, Usman B, Sameer I (2021) From entrepreneurial education to entrepreneurial intention: a sequential mediation of self-efficacy and entrepreneurial attitude. Int J Innov Sci

Chapter 6
Reverse Logistics: A View from Green Capitalism and Circular Economy in Textile Companies in Mexico.

Glenda Marisa Chávez Gallegos

Introduction

Today, and especially in the new economy, the existing explanations for gaining an advantage in the industry have been overturned and new means of gaining an advantage over the competition have been introduced. The fundamental role of concepts such as economies of scale and economies of scope in organizational transcendence are concepts that are part of the past, which gives way to new processes and new ways of doing things, it is at this point where concepts such as circular economy (CE), green economy, green capitalism, and reverse logistics (RL) now take their place in process innovation. It can be explained that just as traditional logistics lost the innovation factor, in the same way, the traditional factors of production, namely land, labor and capital, have lost their role as the main ratifiers of triumph in the business environment of recent days characterized by rapid market changes, proliferation of technologies, fierce competition and product obsolescence.

The competitive landscape has transformed to become a knowledge-based competition and, at the time of today's modernization, organizational precedence lies not only in possessing superior knowledge, but in leveraging it to build superior knowledge-based competences, which in turn are the hallmark of the knowledge economy (Wu 2012). Having stated this, it is extremely important to mention that these changes have been derived from various factors and global changes, marking as the main one the constant search for a model that allows to continue with the current model of life but without the consequences that this model entails, and under this premise relevant concepts such as sustainability and the proposed new model of CE, gives way the entrance of green capitalism as an introduction of the new model of capitalism.

G. M. C. Gallegos (✉)
Universidad Michoacana de San Nicolás de Hidalgo, Morelia, Mexico
e-mail: glenda_c_gallegos@hotmail.com

A. E. Toscano-Hernández et al. (eds.), *Sustainability, Stakeholders and Marketing in the Textile Sector*, SDGs and Textiles, https://doi.org/10.1007/978-981-97-6952-0_6

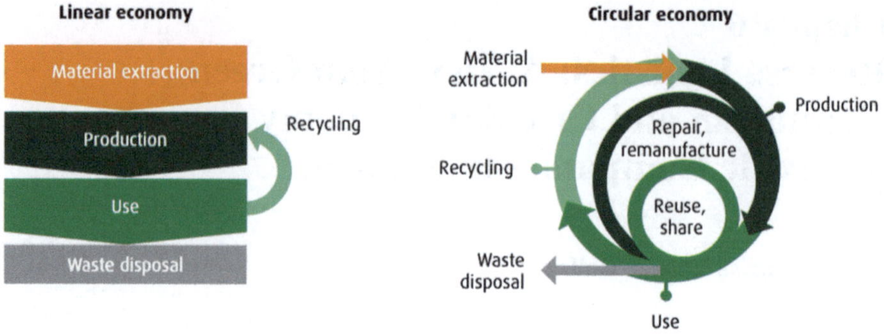

Fig. 6.1 A linear economy and a circular economy. *Source* Own elaboration based on (Stegemann 2023)

A determining event for this conceptualization of new terms was the 1992 Rio Summit, as a landmark event in the study of the environmental variable in the economic and business sphere; it defined the concept of "sustainable development", which was ratified by 180 countries. The "circular economy" model received several denominations among which are: "cradle-to-cradle economy" or "closed-loop economy"; its main proponents insisted that it should not be considered as an ecological movement, but as a way to change global thinking and have a new design philosophy (Macarthur 2013). One of the main exponents Ezio Manzini, proposed sustainable design as a social approach, which took into account factors such as: the environment, culture, production processes, materials, their use, and aspects after their useful life. He also proposed the role of the designer not only as a creator of products, but also of everyday processes, as well as new ideas of well-being (Manzini 2000).

The following figure shows a better example of the comparison between a linear economy, which is the one we currently implement or the current system, and the CE, which is the desired one (Fig. 6.1).

If we analyze this image, we can see that the CE, by emulating nature, is a system that provides durability over time and therefore leads us to sustainability, contrary to the current model.

However, if we focus our attention on the textile sector, it is possible to affirm that some statistics estimate that the textile sector is responsible for about 20% of water pollution, mainly due to dyeing and finishing products. In addition, fertilizers, herbicides, and pesticides are used to grow natural fibers (especially cotton), which penetrate the soil and contaminate aquifers.

Even industrial and domestic washing is not far behind. It is estimated that washing our clothes releases around 0.5 million tons of microfibers into the ocean per year, including many of these synthetic ones (Carrera Gallissà 2017).

At the same time, these microfibers act as a "tax" for chemicals, which when broken down in water only worsen the water pollution problem caused by the textile industry. The type of processes carried out by the textile sector are highly predatory and harmful to the environment, which is why special emphasis is placed on the type of responsibility that these companies should have. This leads to the question: what are these companies doing to implement CE and modify their processes?

Background and Conceptual Elements

In recent years, various academic groups from different disciplines and perspectives of the sciences have begun to actively study possible solutions to the various growing socio-environmental problems (Cone 1980). One of the aspects that has drawn attention in the economic field is the continuous questioning of linear economics and capitalism without social responsibility, emerging multiple types of practices, analyses, and alternatives that will inevitably result in circularity for a sustainable system (Sariatli 2017). Positive factors, theories, and phenomena related to transformative economies can also be distinguished (Sala-Garrido, et al. 2022), which can be a factor that achieves the long-awaited change in social processes and activities, of which companies are an unquestionably important part.

The current production system is mainly based on a linear process of resource consumption, which has been widely criticized for not being sustainable, since it is produced under the "take-make-waste" paradigm. In this system, goods are produced from raw materials, then sold, used and finally discarded as waste, causing not only the depletion of natural resources but also the generation of tons of garbage. About 65 billion tons of raw materials entered the economic system in 2010, and this figure will exceed 82 billion tons by 2020 (Ellen MacArthur 2021).

In the preface to Our Common Future, or better known as Gro Brundtland's report, it is stated that the demand of the United Nations General Assembly for the creation of a "World Commission on Environment and Development" was a clear and forceful demonstration by the international community of the widespread feeling of frustration at the lack of interest or efficiency in responding to the central and decisive global issues. This report became the global reference for the concept of sustainable development, which is understood as development that meets the needs of the present generation without compromising the ability of future generations to meet their own needs. The way of conceptualizing and perceiving this concept has changed a lot over the years, generating two predominant interpretations: the theory of the three sustainabilities, which is considered the best known, and which sees it as an integral concept encompassing, redundantly, three sustainabilities in the social, economic and environmental spheres, and the theory of the dematerialization of growth or decoupling growth from the physical base, which relies, to a large extent, on technology, on technological development as a dematerializing factor and an emerging trend that is highly critical of the previous ones: that of sustainable degrowth (Programa de las Naciones Unidas para el Medio Ambiente 2015).

In this scenario and based on the dematerialization resulting from technology, the circular economy emerges as a valid alternative, which usually implements closed systems and reuse processes, but is much more than that, as it can be conceptualized as a philosophy of "ab initio" design, which in other words can be a restorative or regenerative industrial system inspired by living beings, which emulates the cycles of nature, in which the "waste" of one species becomes the "food" of another and so on, distinguishing between technical and biological cycles. Within this new economy, resources are regenerated within the biological cycle or recovered and restored thanks to the technical cycle (Ellen MacArthur 2021).

To provide a better perspective and explanation, in the following figure it is possible to observe the functioning of the circular economy in an adaptive environment in the industry.

The logic of the CE usually distinguishes between two types of materials or "nutrients" in the economy: biotic and abiotic. Figure 6.2 shows two loop constructs on either side of the product life cycle, going from top to bottom, in what is often called a butterfly diagram. On the left side are biotic materials, also called renewables. They include materials that can decompose naturally, such as food, wood and bio-based plastics. On the other side are abiotic materials, also called non-renewable or finite. These materials cannot decompose, but can be recycled in industrial processes; however, this is not the purpose of the CE, but one of its many alternatives for the use of resources. The basis of the CE is the change of paradigm starting with the design of processes and products.

However, there are several differences between these natural and economic cycles. On one hand, the components of the biological or natural cycle are biodegradable, which means that they can be easily reintroduced into nature; on the other hand, those of the technical cycle are not suitable for returning to nature, since both their design

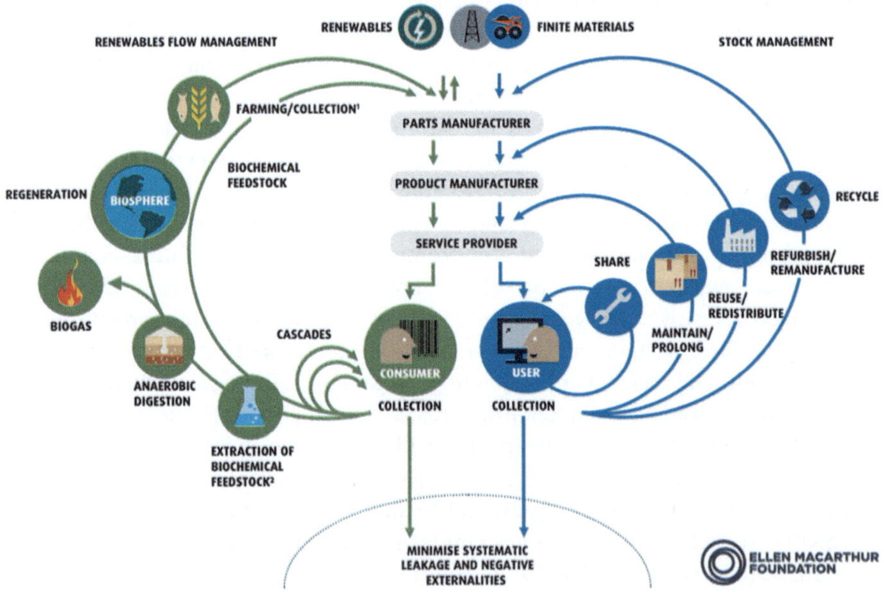

Fig. 6.2 Butterfly diagram of a circular economy. *Source* Figure taken from Ellen MacArthur Foundation, Circular economy system diagram (February 2019)

and components are meant to be assembled and disassembled several times, helping them to be reused again and again until they are returned to the production system. This economy presents a holistic perspective that pursues several objectives much more ambitious than the green economy, as it requires the radical transformation of the means of production, supply chains, as well as the culture and consumption habits of society at a planetary level (Macarthur 2013).

Reverse Logistics

In a CE, products must not only reach the consumer, but also return to retailers or suppliers to enable repair, remanufacturing, and other R's.

Supply chains that go the other way are often referred to as RL. This concept has long existed to enable the return of defective or unsuitable products, however with the conceptualization of the CE it is integrated as a tool for strategic processes in the supply chain thus enabling an eco-friendly design to help achieve a sustainable business.

In a CE, RL is relevant for each and every item that is bought or sold because, at some point, it will have to go to the right place for maintenance, if necessary, and logically subsequent repair, reuse or refurbishment and remanufacturing. Even for recycling, which currently relies on general waste collection, reverse logistics can be important (Stegemann 2023).

It is for this reason that RL represents one of the main tools to align companies to this new economy that pursues closed cycles, closed processes, and that seeks long-term sustainability.

Having a basis for measuring circularity is essential to establish a baseline and assess progress toward a CE. These indicators are often based on a material balance of the system of interest, indicators tend to focus on drivers and responses related to material use. Metrics can also be used at meso and micro scales. In addition, it can be revealing to measure a CE in terms of prevalence of circular strategies (product design, business models, supply chains, reverse logistics, or industrial symbiosis).

Methodology and Research Techniques

This section describes the techniques used and the selection of the sample. First, it was decided to carry out a survey and proceeded to design the measurement instrument, delimiting the companies and characteristics to proceed to apply it. The instrument for both techniques consisted of a questionnaire the first one with 62 questions regarding the RL process and variables such as company, cost, client, product, process, and formalization. The next step consisted of the construction of a table of causes and effects regarding certain dimensions or "variables", such as company, client, supplier, and product. The causes and effects are related directly with the processes in the company and the environmental indicators that affect directly the implementation of RL and that contribute with the achievement of a CE model.

In addition, the list of selected candidates was generated by considering employees of companies that regularly participate in events related to the textile industry and that have RL processes. For the whole process the selection of the people who were meant to be interviewed followed certain characteristics such as being part of the logistics department and process. Participants included 4 or more general distribution managers, 2 vice presidents of logistics, 5 or more returns managers, and 6 or more employees in supervisory positions per company, directly involved with reverse logistics. Strategic positions were selected to ensure that the employees and experts interviewed had the most knowledge of the area and were involved in RL.

The companies that were selected as sampling units for the present study are recognized for being organizations leaders in the field, as well as the main companies with the highest turnover in the textile sector in the country according to various authors (Ammachchi 2017; Akter 2016; Mendonca 2016; Zilly 2016), since they mention that these companies have surpassed all other companies in the last 5 years. The characterization of the sample of companies consisted of choosing companies with high turnover (high volume of sales), with established processes and oriented to reverse logistics, specific to the textile sector that are multinational or transnational and that are physically located in Mexico (Table 6.1).

Once the information collection process was completed, we proceeded to the emptying of matrices for subsequent analysis using SPSS software and fuzzy logic. There are important aspects subsequent to the development of the tool or questionnaire, such as the measurement of validity and reliability.

The validity of a questionnaire is related to the review of the theoretical framework. A good review contributes to the validity of the design of the tool or, in this particular case, of the questionnaire (Rojas Soriano 2006).

Therefore, the reliability and validity of the instrument applied are shown in the following Table 6.2.

Table 6.1 Companies comprising the sample universe

No	Company
1	Liverpool
2	Inditex
3	H&M
4	Grupo Axo

Table 6.2 Confidence score

Cronbach alpha	Cronbach alpha based on standardized items	Number of elements
0.976	0.947	62

Source Own elaboration

Cronbach's alpha quantifies the level of reliability of a measurement scale for the unobservable magnitude constructed from the n observed variables. In this case shows that the questionnaire, which consisted of 62 questions, has validity and therefore the results emanating from it are reliable. In the case of the second tool or contrast tool, the validity lies in the fact that the cause-effect matrix was answered only by experts on the subject.

For the analysis and interpretation of the tool, the following semantic scale was constructed.

Table 6.3 shows a semantic scale based on the work of Gil-Lafuente et al. (2015), in which an evaluation is proposed that includes different levels of LR implementation, ranging from "No influence" (0 on the scale) to "Total influence" (1.0 on the scale); this range of levels. This range of levels makes it possible to identify a difference in the intensity of implementation that helps to identify the relevance of a given variable for understanding the situation of a given variable in understanding the problem situation.

Following that idea, the analysis of the implementation of LR with an effect-cause, cause-cause, and effect-effect perspective was estimated by the experts of the previously selected companies, which have a close relationship and knowledge related to LR, infrastructure and supply chain management. By being aware of possible intermediate impacts or influences, greater internal consistency can be achieved, according to Gento et al. (2001). The results are shown in Table 6.4.

For the purposes of this table, the causes are as follows:

C1: formalization of LR process

C2: adequate infrastructure to execute RL

C3: knowledge of return routes by the customer

C4: knowledge of transformation costs

C5: product design for RL

C6: suppliers involved in the process

Table 6.3 Semantic scale

Level	Semantic label
0.0	No implementation
0.1	Very weak implementation
0.2	Weak implementation
0.3	Low implementation
0.4	Moderated implementation
0.5	Medium implementation
0.6	Considerate implementation
0.7	High implementation
0.8	Strong implementation
0.9	Very strong implementation
1.0	Total implementation

Source Own elaboration based on (Gil-Lafuente et al.2015)

Table 6.4 Matrix cause-effect

	E1	E2	E3	E4
C1	0.8	0.6	0.8	0.4
C2	0.7	0.7	0.8	0.5
C3	0.6	0.5	0.8	0.4
C4	0.9	0.6	0.9	0.7
C5	0.2	0.7	0.7	0.4
C6	0.5	0.8	0.7	0.3
C7	0.7	0.7	0.6	0.6

Source Own elaboration

C7: sustainability policies within the company.
And the effects are:
E1: Waste reduction and consumption
E2: basic human sustenance
E3: Material consumption
E4: Generation and management of waste.

Discussion and Analysis of Results

In this section, the model built is incorporated to show the tool developed from the literature reviewed and the model proposed.

The documentary study shows that the factors affecting reverse logistics are diverse. Likewise, it is observed that there is a certain coincidence among them and that it is possible to group them according to common characteristics, and as a result, blocks of factors are obtained in which the causes that influence reverse logistics in the most decisive terms are concentrated.

The results show those variables that are not fully considered at the time of implementing a reverse logistics process; in that sense, they represent the current state of reverse logistics in the sample companies of the textile sector in Mexico. These results can help, at a national level, the territorial planning and environmental departments of the higher territorial entities of Mexico in the elaboration and dissemination of tools and strategies for environmental activities.

It is important to mention that the theory of forgotten effects allows the researcher to analyze the direct and indirect impacts of the results of the implementation of reverse logistics in companies of the textile sector in Mexico within the framework of sustainable development indicators. With this model it is possible to observe the secondary effects (forgotten effects) that are not directly considered when seeking to implement reverse logistics (Table 6.5).

In relation to these results, Table 6.5 shows that the estimated value of the influence of the actions for the formalization of reverse logistics and the generation and

Table 6.5 Forgotten effects table of influence

Implementation of RL	Sustainable development	M1 C-C	M2 C-E	M3 E-E	M-M
C1: formalization of LR process	E4: Generation and management of waste	0.4	0.8	1	0.9
C3: knowledge of return routes by the customer	E4: Generation and management of waste	0.4	0.7	0.7	0.8
C7: sustainability policies within the company	E4: Generation and management of waste	0.6	0.8	0.9	0.8
C5: product design for RL	E4: Generation and management of waste	0.4	0.8	0.9	0.9

Source Own elaboration

management of materials had a score of 0.4. However, the formalization of reverse logistics has a strong average influence on customer participation, and in turn participation has a strong influence on waste and consumption reduction. Therefore, there are two effects that modified the value of the direct influence on material consumption, resulting in a secondary (forgotten) effect of the mean influence on the formalization of reverse logistics on material generation and management.

With these results, it can be seen that if the process of implementing reverse logistics in companies is formalized, a significant improvement in material consumption indicators can be expected indirectly.

As shown in Fig. 6.3 the implementation of RL has been affected by not considering the formalization process.

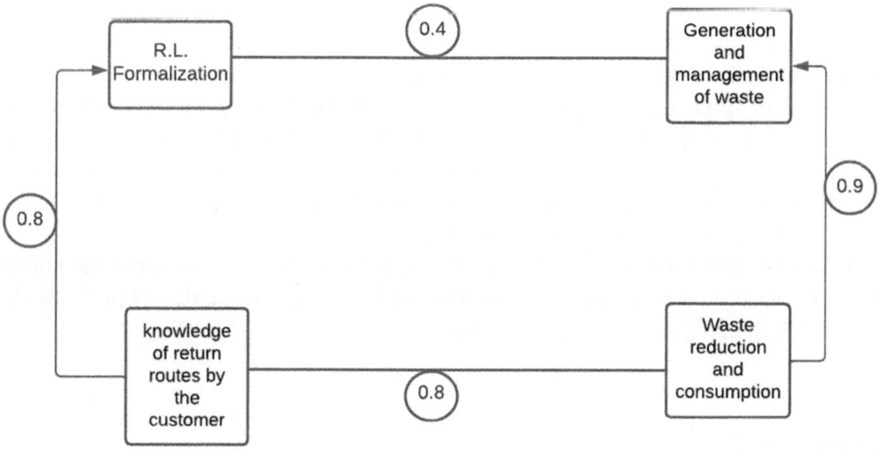

Fig. 6.3 Very strong implementation of RL with certain considerations. *Source* Own elaboration

The results of this study show that using evaluations from experts who have direct operational contact with reverse logistics and who are aware of the application of sustainable development policies, it is possible to identify the components of reverse logistics implementation that affect Mexico's sustainable development. Maximum effects were only identified in the components of formalization and materials management. In order for this sustainable development indicator to be positively impacted, policies aimed at these indicators should be considered to improve the implementation of RL and thus be able to guide companies in the sector toward a CE.

Conclusions

The main objective of this research was to explore and examine reverse logistics in textile companies in Mexico, with a specific focus on the companies that are already implementing it in order to describe and analyze what percentage of implementation and development of their processes they have and with this to determine the feasibility of the use of reverse logistics as a fundamental and basic part in all processes of all companies, and that this concept that is known as innovative is normalized and carried out as a mandatory and daily action in the processes of the companies.

It is undeniable that the results of various sustainable development indexes play a crucial role in shaping public programs and policies aimed at fostering a country's sustainable development, thus improving processes in industry and influencing the judgment of its citizens and preserving its resources. However, when applying the theory of neglected effects to the context of an emerging nation such as Mexico, it becomes evident that certain effects have not been fully considered. This is evidenced by the identification of indicators within these indexes that do not align with observable results in the country's sustainable development.

This work points out the need to conduct a broader study on the influence of sustainable development indicators to develop in order to propose programs and public policies that promote and improve reverse logistics; thus, it is possible to clearly identify how area of formalization of reverse logistics has been an indicator of "forgotten effects" in this area of development.

Likewise, for future lines of research, it is recommended to measure the cause-effect of waste and consumption reduction and to design strategies to promote the five "R's", in addition to reverse logistics.

References

Akter SW (2016) How to improve firm performance using big data analytics capability and business strategy alignment? Int J Prod Econ 182:113–131

Ammachchi N (2017) Ciscto to invest billions in Mexico manufacturing facilities. Obtenido de nearshoreamericas.com. www.nearshoreamericas.com
Carrera Gallissà E (2017) Los retos sostenibilistas del sector textil. Revista De Química e Industria Textil 220:20–32
Cone JD (1980) Environmental problems/behavioral solutions. Cambridge University Press. ISBN: 0521-31973-0
Ellen MacArthur F (2021) Hacia una economía circular. Londres: Ellen MacArthur Foundation
Gammelgaard B (2001) Logistics skills and competencies for supply chain management. Supply Chain Manag 22(2):27–50
Gento AL (2001) Reflexiones acerca de las matrices de incidencia y la recuperación de efectos olvidados. Cuadernos Del CIMBAGE 4:11–27
Gil-Lafuente AM, Santoyo FG, Romero BF (2015) Teoría de los efectos olvidados en la incidencia de la actividad económica en la calidad de vida de los habitantes y cuantificación de los efectos para un reequilibrio territorial. Inceptum 10(19):105–122
Macarthur E (2013) Report:Towards the Circular Economy. Recuperado de, http://www.ellenmacarthurfoundation.org/circular-economy/circular-economy/the-circular-model-an-overview
Manzini EB (2000) Ecología y democracia, Barcelona. Editorial Icaria. ISBN: 978-84-7426-497-5
Mendonca J (2016) IT companies including TCS, Infosys, Wipro bracing for slowest topline expansion on annual basis
Programa de las Naciones Unidas para el Medio Ambiente (2015) Descubriendo los caminos hacia una economía verde e inclusiva. Ginebra: Resumen ejecutivo Programa de las Naciones Unidas para el Medio Ambiente
Rojas Soriano RJ (2006) Guía para realizar investigación social
Sala-Garrido R, Mocholí-Arce M, Molinos-Senante M, Maziotis A (2022) Measuring operational and quality adjusted efficiency of Chilean water companies. NPJ Clean Water 5(1):1. https://doi.org/10.1038/s41545-021-00146-x
Sariatli F (2017) Linear economy versus circular economy: a comparative and analyzer study for optimization of economy for sustainability. Visegr J Bioeconomy Sustain Dev 6(1):31–34. https://https://doi.org/10.1515/vjbsd-2017-0005
Stegemann SV (2023) An introduction to waste management and circular economy. UCL Press, London
Wu I (2012) Examining knowledge management enabled performance for hospital professionals: a dynamic capability view and the mediating role of process capability. J Assoc Inf Syst 13(12):976–999
Zilly JG (2016) Recurrent highway networks. In: International conference on machine learning. Recurrent highway networks, pp 4189–4198

Chapter 7
Green Marketing Strategies in Sustainable Fashion Business Models

Leydis Ester Córdoba Pastrana and Johana Fajardo

Introduction

The use of sustainability terminology within the context of the fashion system began at the beginning of the 21st century (Martínez-Barreiro 2020), but prior to this terminology, the concept of fast fashion was developed in 1950 to meet mass and seasonal production (Castro 2021), which has required companies to optimize their supply chain to respond quickly to new fashion trends and consumer demands by putting fashion products in stores with a delivery time of less than 1 month.

Contrary to fast fashion, the concept of slow fashion or slow fashion is emerging, defined by the Spanish association of sustainable fashion as "any garment or accessory that respects the environment, human and workers' health, and promotes the use of sustainable materials, the reuse of existing materials and local production". This evolution in the fashion industry is an opportunity for a major transformation towards sustainability.

In Colombia, the textile sector constitutes one of the main sectors of the economy in its market size in the first quarter of 2023 reached 7.52 trillion pesos, 5.8% more than the same period of 2022. In percentage values, it represents 3.1% of consumer spending in Colombia (Figures: Raddar, Elaboration: Sectorial).

According to the Fashion Observatory 2023, this consumption is represented in 50.7% by the apparel category, 21.1% by personal items and accessories, 11.5% by footwear, 9% by clothing and footwear services, and 7% by household linen, and according to the Dane technical bulletin, in the 2020 industrial environmental survey,

L. E. C. Pastrana (✉)
DC Consultorías SAS, Colombia, Colombia
e-mail: gerencia@dcconsultorias.com

J. Fajardo
Universidad Cooperativa de Colombia, Colombia, Colombia
e-mail: Johana.Fajardope@campusucc.edu.co

A. E. Toscano-Hernández et al. (eds.), *Sustainability, Stakeholders and Marketing in the Textile Sector*, SDGs and Textiles, https://doi.org/10.1007/978-981-97-6952-0_7

in Colombia, the textile industry generated 76,298 tons of solid waste and disposed of 32,174 tons.

According to Procolombia's 2023 report, the Colombian textile industry is characterized by five competitive factors: innovation and diversification, experience and reliability, the country's strategic location and commercial advantage, competitive prices, quality raw materials, and commitment to sustainability.

This last factor makes sustainable fashion companies increasingly interested in transforming their processes towards business models focused on sustainability and commitment to the Sustainable Development Goals (SDGs).

The marked growth of consumption in the sector raises questions about the way companies develop their processes and approach consumption, as well as sustainability policies with regard to processes, products, people, and the environment.

However, fast fashion and slow fashion trends mark the segmentation of consumption in this sector. It is notorious that concepts such as fast fashion are fundamentally based on the establishment of low prices, diversity, and high rotation of collections (Sánchez-Vázquez et al. 2020). On the other hand, slow fashion shifts the focus towards quality improvement, increasing the wear time of garments, with an eye towards sustainability and having a defined social and environmental approach.

Research evidences how mainstream consumers feel about and choose sustainable fashion, with environmental concerns and social norms being the main reasons why people buy sustainable fashion (Kim et al. 2020).

Given the growing need to adopt measures to benefit the environment, awareness and care of resources and consumer demands, it is relevant to analyze how companies that have adopted this sustainable business model based on slow fashion have established marketing strategies that allow them to publicize their products, capture the attention of the end consumer and, at the same time, raise awareness of the importance of their value proposition as a sustainable business, encouraging consumers to contribute to the care and protection of the environment.

The present research dives into the detailed analysis of green marketing strategies employed by sustainable fashion companies, exploring audience interaction with these strategies and their impact on end-consumer engagement with sustainability and the SDGs.

Literature Review

Sustainable Fashion is a subset of the slow fashion movement, and the terms eco-, green-, and ethical fashion are generally used synonymously (Carey & Cervellon 2014).

The notion of sustainability in the fashion industry became popular after the 1992 Rio Earth Summit, when brands began to explore ways to reduce their negative impact on the planet and communities. Sustainable fashion (SF) not only embraces sustainability in every phase of its life cycle, thus mitigating its environmental impact and promoting social responsibility, but also focuses on designing, creating, and marketing socially and environmentally responsible products (Ray & Nayak 2023).

The essence of sustainable fashion (SF) lies in addressing the fundamental problem of rapid obsolescence and early disposal that garment design and production are currently geared towards. This approach, focused on generating quick revenue, contributes to making the textile, clothing, and fashion (TRM) industries some of the most polluting in the world, generating an alarming volume of garment waste (Grazzini et al. 2021).

TRM industries are not only responsible for eight to ten percent of global greenhouse gas emissions, but also play a central role in global warming. This phenomenon has palpable consequences in everyday life, manifesting itself in extreme weather events such as floods, droughts, and hurricanes, along with sea level rise (Shrivastava et al. 2021; Mishra et al. 2021). To address these challenges, 195 countries joined together in the Paris Agreement in 2015, committing to limiting temperature rise to less than 2 °C (Doukas et al. 2018).

Climate change arises as a negative consequence of unsustainable human practices, exacerbated by the industrial revolution, technological disruption, and automation. This intrinsic relationship between climate change and business is reflected in the physical risks this phenomenon imposes on supply chains, business models, and decision-making processes (Woetzel & Krishnan 2020). Governmental regulatory pressures continue to drive substantial changes in technological, material, organizational, economic, and socio-cultural aspects of business (Kivimaa et al. 2019).

In this context, the United Nations launched the Sustainable Development Goals (SDGs) in 2015, with 17 goals and 169 targets, urging countries to prioritize environmental sustainability, social inclusion, and economic development (United Nations 2015). To achieve these SDGs, fashion companies must intensify their commitment to corporate social responsibility (CSR) in diverse areas, promoting more sustainable production and consumption patterns, reducing water consumption, and ensuring decent working conditions (Sauermann et al. 2020).

Following these concerns and in response to the growing interest in the UN SDGs, TRM industries have shown an increased propensity in recent years to address sustainability issues (Kabir et al. 2019). Growing environmental awareness has engaged both businesses and consumers in actions aimed at global collective benefit, leading to stricter environmental regulations that force entrepreneurs to be more aware of the environmental impacts of their actions (Castellano et al. 2015).

This business commitment to sustainability has driven the emergence of marketing practices, known as green marketing, that seek to promote environmentally friendly products and understand consumers, their habits, and purchasing patterns (Aguilar 2016). Companies in TRM industries are adopting initiatives that embrace eco-efficiency concepts and seek to implement environmental practices, including sustainability reporting (Muñoz-Torres et al. 2021).

The American Marketing Association (AMA) defines green marketing from three perspectives: commercial, social, and environmental. This strategic approach addresses the promotion, production, recovery, and packaging of goods in a responsible manner and in line with environmental concerns (Novillo Díaz 2018). The challenge of green marketing lies in developing strategies that not only drive awareness and commercialization of sustainable products, but also modify consumer purchasing behavior, thus expanding the segment of green consumers (Chamorro 2001; Alí García-Capdevilla et al. 2021).

In addition, marketing plays an essential role in achieving the SDGs by facilitating responsible consumerism and sustainable business practices (Voola-Azmat 2022). However, in the implementation of these strategies, the emergence of greenwashing practices is observed, the purpose of which is to mislead consumers about the environmental practices of an organization or the environmental benefits of a product or service (Delmas and Burbano 2011, Marko & Kusá 2023).

Authentic green marketing involves the real integration of sustainability in all company processes, from the design of products that meet consumer needs with the minimum use of resources to the reduction of environmental impact in distribution and the promotion of an efficient reverse distribution system (Chamorro 2001; Véliz & Freire 2019). Green communication, through various channels, plays a crucial role in informing consumers about the environmental features and benefits of products, while conveying the image of a company committed and active in environmental issues (Calomarde 2000, Evans & Peirson-Smith 2018).

The key strategies of green marketing include eco-design, adoption of environmental certifications and labels, and promotion of sustainable innovation). These approaches, integrated into the corporate culture, affect the purchase decision, brand positioning, and consumer recall (Martinez-Rodriguez et al. 2018).

Likewise, within the conclusions made by the studies, marketing strategies and public communication campaigns should focus on increasing environmental awareness and the perceived value of sustainable products to positively influence consumer behavior.

According to Ray and Nayak (2023), SM marketing can be analyzed in two main ways depending on the end users: "Business to Consumer (B2C)" and "Business to Business (B2B)". In the context of fashion, where the predominant brand focuses on the B2C market, both in physical and online commerce, it faces constant challenges of competition and trends. However, sustainable fashion is often limited by higher prices and lower accessibility, which can hinder its mass adoption (Spielmann 2021). To counter this, brands are incorporating effective marketing strategies that emphasize sustainable awareness to expand their business reach (Kong et al. 2016).

The theory of planned behavior sheds light on the purchase intention of sustainable fashion, evidencing positive correlations with factors such as ecological trust, environmental awareness, and consumer attitude, which underlines the complexity of the determinants in this context (Dhir et al. 2021; Hwang et al. 2020).

The fashion industry has explored various strategies to implement sustainable business models, including high-quality products, seasonal products, second-hand/used garments, repair and customization services, fair trade, rental, and exchange

(Pal & Gander 2018). As well as, the adoption of organic, natural, and biodegradable fabrics, recycled fibers, non-toxic colors, zero-waste design, plastic-free packaging, artisanal techniques, and limited collections for various seasons are key considerations for sustainable fashion business models (Pal 2017).

Another essential component in this application of marketing strategies to expand its commercial reach is technology, which can contribute to sustainable solutions for the fashion industry (Boons & Lüdeke-Freund 2013). The application of cutting-edge technologies allows us to discover new opportunities for the production of sustainable raw materials and the recycling of used clothing (Olatubosun & Bello 2024).

Therefore, an example of the adoption of technological advances is the integration of 3D virtual sampling business models for the design, purchase, and sale of clothing, which is transforming the traditional cycle that requires physical sampling. 3D virtual design significantly reduces material, energy, and transportation waste, accelerating the design and production process by allowing detailed visualization and review without the need for multiple physical samples (Pal 2017; Salamzadeh et al. 2023).

At the same time, there are body scanning business models that overcome the limitation of standard sizes in clothing production. Sustainable business models incorporate body scanning to ensure personalized clothing tailored to consumers' exact measurements, avoiding returns, and reduce waste (Berger-Grabner 2018).

On the other hand, there are Alternative Textile Business Models, which, although they do not implement technologies, innovate in the model, given the unsustainable demand for resources in the production of conventional fibers and textiles, which explore alternative ecological fibers such as hemp, bamboo, ramie, and innovative materials such as Pinatex and Muskin leather, derived from pineapple leaves and fungi, respectively (Pandit et al. 2020).

Finally, on-demand fashion business models, which are driven by demand, facilitated by e-commerce, emerge as an effective strategy to reduce material waste and minimize overproduction. Manufacturers can receive virtual orders and adjust their production accordingly, adapting to specific customer needs (Caetano 2022; Christen 2018).

In congruence with the above, this research seeks to offer a more complete and contextualized understanding of the online dynamics related to sustainable fashion, thus contributing to filling possible knowledge gaps in the field. Furthermore, when analyzing the green marketing strategies implemented by sustainable fashion companies on digital platforms. Therefore, the methodology is presented as a bridge between theory and practice that promises to enrich the existing literature.

Methodology

The netnographic study is presented as a qualitative and interpretive approach specifically designed to explore consumer behavior in the environments of communities and cultures in use of the Internet (Gebera 2008). Netnography uses publicly available information from forums and social networks to identify and understand the needs and factors that influence the decisions of consumer groups (Kozinets 2019).

In this research, netnographic data collection was conducted through immersion and active research (Kozinets 2019). Immersion involves the researcher inhabiting the conceptual space of the project, with the objective of identifying green marketing strategies applied by companies on social networks and analyzing audience and media response.

The immersion process included exploring, selecting, recording, and reflecting on data from various social media platforms and websites. A lexicon of keywords and phrases was developed to track relevant online sources. Thus, Colombian governmental online sources were examined, where companies in the textile sector are presented under the sustainable fashion business model, along with news articles that enriched the researcher's understanding of nationally recognized sustainable fashion companies with a digital presence. This diversity of sources allowed for obtaining a deeper understanding of the application of sustainable marketing strategies toward the consumer through social networks.

In the development of the research, a purposive sampling was used to identify different types of companies with a sustainable fashion business model. Since the focus is on analyzing green marketing strategies, the main selection criterion was to choose companies with an active presence in social networks, which are one of the communication channels where these companies deploy their green marketing strategies to consumers or potential customers and in which the audience's behavior can be analyzed.

The social media immersion focused on exploring the accounts and posts of sustainable fashion companies, analyzing images, descriptions, captions and comments from followers.

The research was carried out in 20 companies in the sector studied, analyzing their social networks for 6 months, and examining their content and user comments on a daily basis. The selection of companies was based primarily on information obtained from the online sources mentioned above, using the lexicon of keywords and phrases associated with sustainable fashion and the SDGs. This provided a general understanding of how companies under the sustainable fashion business model interact with users on social networks to deploy their marketing strategies. In addition, a variety of complementary data, including audio, videos, and images, were analyzed to better understand the phenomenon under study.

Three filters were applied to select the companies. First, the companies' social network accounts had to be active, i.e., with at least one monthly publication; second, the accounts had to be at least 1 year old. Thirdly, the content posted on the social network had to be associated with topics related to their business, not with isolated topics or those that had no relation or connection with the company's own activity, since the objective of the research was to explore how the audience interacted with the publications used as part of the green marketing strategy.

Table 7.1 shows the brands selected, and the total number of followers in the social networks that each brand has, which added together give the total audience available for each company.

Table 7.1 shows that the social network with the largest audience is Instagram, a fact that determines its selection as the main source of data collection.

To gain a deeper understanding of the contextualized data, the paid data mining tools RivalIQ and SocialElephants were used to collect information related to posts, follower interaction with posts, social media activity.

In Fig. 7.1, it is evident that the most active channel is Instagram. The channel that generates the highest participation in its panorama is TikTok, likewise, the channel's publications per day by participation rate per follower of the company and Instagram

Table 7.1 Selected Brands and Total Audience

Brand	Facebook followers	Instagram followers	Followers TikTok	YouTube followers	Total audience
Lafayette	72,000	55,400	0	4170	131,570
Sixxta	1200	121,000	0	0	122,200
Cyclical	1100	100,000	5517	0	106,617
Waimari	0	102,000	0	0	102,000
The relicka	0	82,500	0	0	82,500
Garage sale by EM	728	56,800	0	6260	63,788
Musth have	0	59,400	0	0	59,400
Fabricato	21,000	16,400	0	434	37,834
Gypsian acollective	433	32,700	0	0	33,133
Aramacao	0	27,700	0	0	27,700
Serial resort	0	25,500	124	0	25,624
New the label	0	19,500	0	0	19,500
Nawa	0	19,100	0	0	19,100
Itana	0	16,100	0	0	16,100
Cintoconce pt	36	14,100	0	0	14,136
Cult Hera	0	10,200	2	0	10,202
ephedra.co	0	9033	0	0	9033
On My Way	0	2414	30	0	2444
Cicreytex	0	1528	0	0	1528
Osse.co	0	1369	0	0	1369

Source Social networks and own elaboration (2024)

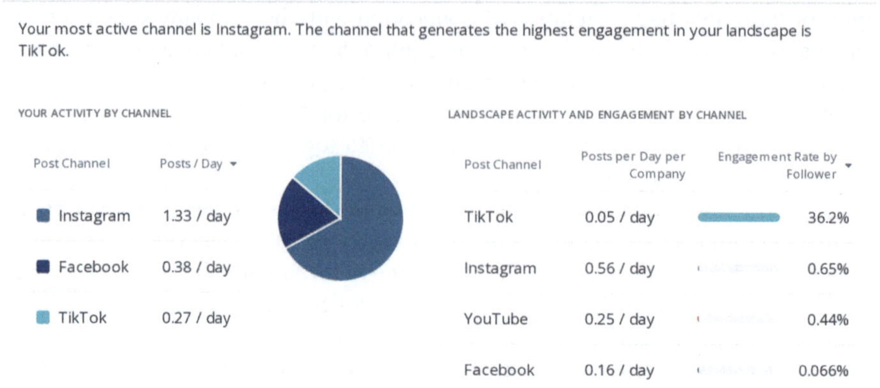

Fig. 7.1 Post Channels. *Source* Social networks and own elaboration (2024)

1.33/day ‘TikTok 0.05/day, 36, 2% EB Facebook 0.38/day Instagram 0.56/day 0.65% and TikTok 0.27/day YouTube 0.25/day 0.44%.

This approach allowed for a high level of netnographic sensitivity by collecting a voluminous amount of relevant data and purposefully analyzing a smaller subset. The subset of posts with the most “Likes” and most comments was identified using data mining tools (Reid & Duffy 2018). These posts were sampled because posts with a higher number of “Likes” and comments were more likely to provide a good combination of source and content factors. Sampling was limited to static posts, as Instagram with 87% of the total audience, is the social network with the highest number of followers for the sustainable fashion business model companies sampled, is primarily used for photo sharing, and it was observed that static posts were the most common and significant form of content that users interacted with during the immersion stage (Xie-Carson et al. 2023). The data research phase lasted six months, from June 01, 2023 to December 01, 2023.

After collecting all posts on Instagram accounts during the research period, the URL of each post was used to collect user comments using the paid comment extraction tool https://exportcomments.com/. These posts were exported to Excel spreadsheets, including details such as social network name, date, number of “Likes” and comments. The posts were then sorted in ascending order according to the number of “Likes”, and the comments that received the most “Likes” were collected until thematic saturation was reached. This sampling strategy was used because the comments that received the most “Likes” were popular and identifiable among users, indicating that they were more representative of the opinions of users in general. Likewise, these publications were segmented depending on the disclosure approach given by the brand towards the main SDGs associated with the industry and in which the company is focused on its compliance.

User engagement in social networks manifests itself at three levels: consumption, contribution, and creation (Schivinski et al. 2016). Consumption involves passive activities such as viewing images and videos, as well as following a topic online. Contribution encompasses active participation, such as liking, sharing, or commenting on social network posts. Creation represents the highest level of engagement and includes actions such as posting user-generated content (Schivinski et al. 2016).

The research analyzes user engagement at the levels of consumption and contribution to the content generated by sustainable fashion brands within the framework of green marketing strategies. In order to maintain a manageable sample size, the five (5) static posts that received the most "Likes" and most comments from each of the 20 companies were collected for further analysis. The final dataset incorporated a total of 100 Instagram posts and 650 user comments associated with these posts.

Data Analysis

Adequately describing a communication process involves asking fundamental questions such as: Who says what, in what channel, to whom and with what effect? These questions guide the fundamental lines of communicative analysis, making part of the coding process: the source and the content of the communication process (Herrera et al. 2021).

In the research, prior to coding, user comments on the collected posts were iteratively examined to gain a general understanding of the data. During this phase, a third category was identified that intertwines source factors and meaningful content: the parasocial relationship factor. This factor represents the connection developed between users and influencers (entrepreneurs, business people, or content creators in companies) through the exposure of their content on social networks (Xie-Carson et al. 2023). Therefore, the parasocial factor constitutes an intertwining of source and content factors, as it is closely associated with both the influencers employed by brands and their posts.

The concept of 'parasocial relationship' refers to the one-sided (i.e., non-reciprocal), close, long-term socioemotional relationship or bond that the individual illusorily forms with the media personality through successive exposures (Cohen 2004).

The netnographic analysis followed an inductive approach, using images and captions within these publications as context to interpret the comments of the 100 selected publications.

In the first round of thematic coding, Microsoft Excel was used to list the 100 Instagram posts collected during the data search process. Each Excel sheet detailed the Instagram images, title, and comments for each post. Comments that did not contain text, did not make sense, or were not relevant to the source or content factors were removed from the analysis. In the second round of thematic coding, Microsoft Word was used to form categories using codes identified in the Excel sheets. The codes were generated, evaluated, and grouped into common categories following an iterative coding and retrieval process. During this process, a number of theoretical constructs from studies on green marketing strategies and sustainability were considered as potential categories for coding.

In the final round of coding, using the Atlas ti tool, the relationships between categories and themes associated with the identified categories were examined. For the qualitative analysis, taking into account the information from the previously selected publications and the comments extracted from the social networks, the marketing strategies employed by the sustainable fashion businesses were identified and analyzed. In addition, the reaction or comments of the final or potential consumer, called social network user, to the green marketing strategy developed were analyzed, addressing issues related to the sustainability of the products, the use and relevance of sustainable fashion, the environmental impact and social responsibility of the brand, as well as community education on the importance of environmental care and protection and how the brand contributes to the fulfillment of the Sustainable Development Goals (SDGs).

Results and Discussion

First, an analysis of the characteristics of the sustainable fashion companies considered as a sample was carried out, followed by an analysis and description of the green marketing strategies employed by the brands and the behavior of consumers in relation to these strategies.

The quantitative analysis of the 20 sustainable fashion brands considered as a sample shows that the 5 brands with the largest audience on social networks are: Lafayette, an iconic brand with 80 years in the Colombian textile industry, located in Bogota, has a community of over 131,570 followers, followed by Sixxta company created in Girón, Santander, with 122,200 followers. Cíclico, a company from Bogota, with 106,617 followers has the highest growth in followers and a 74% growth in its audience during the evaluation period, followed by Waimari, a company from Barranquilla with 102,000 followers, and The Relicka, a company from Bogota with 82,500 followers. This information in detail is provided in Table 7.2.

Audience growth for these brands is observed for each of the brands during the period from June to December 2024, with 9 of the 20 brands showing double-digit audience growth.

Table 7.3 shows the total audience and the average number of monthly posts on Instagram, which is the network where 87% of the followers for these brands are located, and the average engagement or interaction of the audience with the posts.

The data obtained in Tables 7.2 and 7.3 allow inferring that brands with a greater number of followers and a greater number of publications have achieved a greater increase in their audience than brands with a lower frequency of publication.

Analysis of SDGs in sustainable fashion brands.

According to Calvo (2018), sustainable fashion companies are characterized by their objective of reducing the negative impact of the industry on the environment and

Table 7.2 Brands and total audience in social networks

Brand	Total audience 01-06-2023	Total audience 01-12-2023	% Audience growth	Average monthly Instagram posts
Lafayette	123,023	131,570	7%	9
Sixxta	102,006	122,200	20%	20
Cyclical	61,396	106,617	74%	48
Waimari	85,000	102,000	20%	22
The relicka	57,006	82,500	45%	20
Garage sale by EM	52,394	63,788	22%	16
Musth have	45,670	59,400	30%	50
Fabricato	36,217	37,834	4%	20
Gypsian A collective	32,280	33,133	3%	12
Aramacao	24,000	27,700	15%	14
Serial resort	23,000	25,624	11%	34
New the label	19,012	19,500	3%	10
Nawa	18,776	19,100	2%	10
Itana	15,674	16,100	3%	9
Cintoconcept	13,467	14,136	5%	10
Cult Hera	8798	10,202	16%	7
Ephedra.co	8560	9033	6%	12
On My Way	2400	2444	2%	2
Cicreytex	1450	1528	5%	5
Osse.co	1280	1369	7%	7

Source Social networks and own elaboration (2024)

society, thus contributing to the achievement of the Sustainable Development Goals (SDGs). These include ensuring the availability and sustainable management of water and sanitation for all (SDG 6), ensuring access to affordable, secure, sustainable, and modern energy for all (SDG 7), building resilient infrastructure, promoting inclusive and sustainable industrialization, foster innovation (SDG 9), make cities and human settlements inclusive, safe, resilient and sustainable (SDG 11), and ensure sustainable consumption and production patterns (SDG 12).

The 20 selected companies are framed within alternative on-demand textile and fashion business models, for which, taking into account the SDGs, the themes of the publications of each of the selected brands are analyzed, and the publications are classified according to the sustainable development goal. The brand and through which the company focuses its green marketing strategy, promoting communication through publications to its audience and making known how the brand is contributing to compliance with the SDGs and making its individual work known to them.

Table 7.3 Number of monthly publications and Engagement %

Brand	Total audience 01-12-2023	Average monthly publication s	Average like per post	Average comments per post	Average engageme nt
Lafayette	131,570	9	161	6	0.13%
Sixxta	122,200	20	609	9	0.51%
Cyclical	106.617	48	402	26	0.42%
Waimari	102.000	22	3	4	0.02%
The relicka	82,500	20	1319	35	1.62%
Garage sale by EM	63,788	16	138	4	0.24%
Musth have	59,400	50	24	1	0.12%
Fabricato	37,834	20	100	5	0.32%
Gypsian A collective	33,133	12	4	1	0.05%
Aramacao	27,700	14	124	7	0.50%
Serial resort	25,624	34	63	3	0.38%
New the label	19,500	10	80	4	0.46%
Nawa	19,100	10	3	1	0.07%
Itana	16,100	9	3	5	0.07%
Cinto concept	14,136	10	115	17	0.88%
Cult Hera	10,202	7	92	2	0.97%
Ephedra.co	9033	12	3	1	0.17%
On My Way	2444	2	21	1	0.94%
Cicreytex	1528	5	21	0	1.70%
Osse.co	1369	7	12	4	1.39%

Source Social networks and own elaboration (2024)

Table 7.4, evidences the classification given to the 100 Instagram posts selected (5 for each brand), which were segmented according to the focus of work by the company towards the fulfillment of the SDG intended to be communicated to the audience.

Table 7.4 shows the focus given by the brands to the publications. Of the 20 companies sampled, 16 of them have publications with the highest number of likes and comments oriented towards their commitment to SDG 9 and 12.

Of the 100 publications with the highest interaction among the audience, 36% is directed towards sustainability issues and the remaining 64% is directed toward marketing strategies where brands directly promote their products to lead the audience to purchase.

Identification and analysis of green marketing strategies.

Table 7.4 SDGs promoted in the publications

Brand	SDG 9	SDG 12	Other commercial brand objectives
Lafayette	40%	40%	20%
Sixxta	20%	20%	60%
Cyclical	40%	40%	20%
Waimari	–	40%	60%
The relicka	20%	20%	60%
Garage sale by EM	20%	20%	60%
Musth have	0%	0%	100%
Fabricato	40%	60%	0%
Gypsian A collective	0	0	100%
Aramacao	0	20%	80%
Serial resort	0	0	100%
New the label	0	0	100%
Nawa	0	20%	80%
Itana	0	20%	80%
Cintoconcept	20%	20%	60%
Cult Hera	0	20%	80%
Ephedra.co	20%	40%	40%
On My Way	20%	40%	40%
Cicreytex	20%	40%	40%
Osse.co	0%	0%	100%

Source Social networks and own elaboration (2024)

After analyzing the publications that are directly related to the topics associated with sustainability and SDGs, we proceeded to identify the green marketing strategies applied by these companies with a sustainable business model focused on the B2B market:

Strategy 1: Education and audience awareness.

In the brands analyzed, it can be seen how this strategy is the one most used by brands when communicating with their audience, 25% of the brands are oriented toward informing and educating consumers on issues associated with the concept of sustainability, responsible consumption, the importance of the SDGs and how to preserve the environment. The implementation of this strategy is evidenced in 42% of the publications oriented towards topics associated with sustainability.

A consumer committed to environmental policies demands two types of information: scientific knowledge about the relationship between productive activities and the environment; and, secondly, information related to the production process itself and its impact on the environment. Green marketing strategies are used by companies worldwide (Burguillo & Jurjo 2007).

The Cyclic brand, for example, engages its community in its publications by raising awareness of the meaning of concepts such as sustainability, carbon footprint, and the possible uses of textile waste and how its products can be reused to give them a new life cycle, in addition to fostering a ‘parasocial relationship’ with its audience to achieve a greater connection with the publications.

Strategy 2: Environmental labeling:

Ecolabeling is an environmental policy instrument capable of attracting consumers who are concerned about the impact of their consumption habits on the environment and who are willing to change their consumption habits by switching to products with a lower environmental impact, provided that they provide truthful information on the subject (Burguillo & Jurjo 2007).

Eco-labels are considered an exceptional marketing and advertising tool that informs contemporary consumers about the ecological characteristics of a product (Martinez-Rodriguez et al. 2018).

As part of this strategy, the Lafayette brand explains in its publications on social networks, that each of the sustainability seals that the fabrics of its products have: Better Cotton, The organic Content Standard, and Global Recycled Standard, also invites its community to explore to investigate about the different existing sustainability seals globally and share them to the community of followers.

Within the total number of companies selected, 10% of the companies use this strategy: Lafayette and Cíclico. 3% of the publications focused on sustainability are part of the deployment of this strategy.

Strategy 3: Product innovation and eco-design:

According to the Spanish Chamber of Commerce, eco-innovation seeks to optimize all phases of production through the modification of production and consumption patterns, seeking to make efficient use of natural resources by analyzing their life cycle, as well as the application of new business management methods that avoid or minimize environmental damage.

Product eco-design strategies involve: the development of new concepts, selection of low-impact materials, reduction of material use, optimization of production techniques, optimization of the distribution system, reduction of impact during use, optimization of the initial service life, and optimization of the system at the end of its service life (Obrecht et al. 2019).

In 55% of the publications associated with sustainability issues, the strategy is to publicize new products created from sustainable ideas and materials, such as the use of PET, various recyclable materials, and handmade production, for which the brands, in addition to communicating product ideas, involve followers by testing the type of product they would like the brand to launch in the market, and encourage the audience to share the news of similar eco-friendly products they have seen in other countries. This strategy is applied in 35% of the companies analyzed, being the one with the greatest impact and acceptance in the analyzed business model companies and their audience.

Strategy 4: Sustainable collaborations and alliances:

Partnerships between companies and NGOs are becoming an increasingly important tool for advancing the sustainable development agenda. The strategy focuses

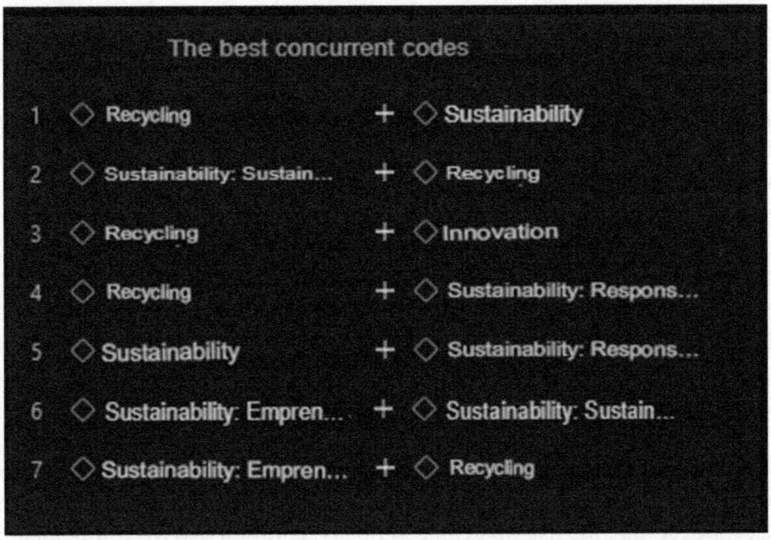

Fig. 7.2 Green marketing strategies for the final consumer. *Source* Social networks and own elaboration (2024)

on communicating to current and potential consumers the partnerships established with brands that share the goal of supporting environmental conservation (Nielsen 2019). In this approach, a message of togetherness towards corporate social responsibility is conveyed, in addition to engaging communities to raise awareness about environmental care.

When analyzing this strategy in the sample of sustainable fashion companies, the brands highlight their efforts in the optimization of processes, evidencing in a transversal way joint collaborations to promote the development of eco-friendly products. In addition, related brands that are aligned with the principles of sustainability are made known.

It is important to note that this strategy is only implemented in 10% of the brands taken as a sample and in 3% of the selected publications with a focus on sustainability, being Lafayette and Cíclico exceptional examples, since in the selected publications they make known to the audience their alliances with brands such as Tatacoa Race, Errata Natura and Americanino. The remaining 99% of the brands do not make use of this green marketing strategy (Fig. 7.2).

Impact of Green Marketing Strategies on the End Consumer

Taking into account the identified marketing strategies used by these brands to publicize their commitment to the SDGs, we proceeded to analyze the behavior of followers in the different publications aligned with said strategies, for which 650 comments from the brands were studied. 100 publications were made in the analysis period.

In turn, concurrent themes were identified that represent the communication and interaction between brand and consumer, as evidenced in Fig. 7.1. Sustainability + Recycling, this combination of terms integrates sustainability and recycling into the commercial practices of fashion brands Sustainable. To include a holistic vision that includes the reduction of environmental impact throughout the entire supply chain, from production to consumption and eventual recycling of products, as an essential component of this approach, where materials and products are designed and manufactured in a way that they can be reused or recycled at the end of their useful life, thus contributing to the conservation of resources and the reduction of waste.

Secondly, Recycling + Innovation highlights the importance of innovation in the field of recycling within the sustainable fashion industry. The brands are committed to developing and adopting new technologies, processes, and creative approaches that improve the efficiency and effectiveness of recycling textile materials. This may include, for example, the use of advanced fiber separation and recycling techniques, the implementation of reverse logistics systems to facilitate the collection and recycling of used clothing, or the exploration of new materials and manufacturing methods that minimize the environmental impact.

Thirdly, Recycling + Sustainability: Social Responsibility highlights the brands' social commitment to recycling and sustainability. In addition to adopting sustainable practices in their operations, these brands also care about the social impact of their actions. This may involve supporting local communities through recycling and employment programs, promoting environmental education, and raising public awareness about the importance of recycling for social and environmental well-being.

Fourthly, Sustainability + Social Responsibility shows the connection between sustainability and social responsibility in general. The brands that promote this combination are committed to operating in an ethical and responsible manner, both in environmental and social terms. This may include respecting labor and human rights throughout the supply chain, adopting transparent and ethical business practices, and committing to relevant social causes such as environmental justice and gender equality.

Fifth, Sustainability: Entrepreneurship + Sustainability: Sustainability, this combination underlines the role of entrepreneurship in promoting long-term sustainability. Here, brands are characterized by their strong business approach and continued commitment to sustainability in all areas of their business. This may include constantly seeking new opportunities to innovate and improve environmentally and socially, as well as adopting business management practices that foster sustainability over time.

Finally, Sustainability: Entrepreneurship + Recycling: In this combination, the entrepreneurial potential is highlighted to address the challenges of recycling and sustainability in the fashion industry. Here, brands are committed to using their entrepreneurial spirit and innovation capacity to find creative and effective solutions to problems related to the recycling of textile materials. This may involve creating circular business models that encourage the reuse and recycling of products, as well as collaborating with other industry players to promote systemic change towards more sustainable practices.

These results show the alignment of the responses of the social network community with the green marketing strategies implemented, and in them, we can identify how the comments of potential consumers are focused on an interest in greater knowledge about this type of brands and their sustainability initiatives, the support they give to brands of complementary products or of different categories with similar sustainability objectives and the great interest of users to learn about the use of waste textile products, such as reuse of fabric scraps.

Challenges with sustainability practices in the textile sector.

The challenges of sustainability practices in the textile sector present both obstacles and opportunities for companies.

Among the challenges identified in the companies are those related to brand differentiation, specifically those related to the construction of a solid brand identity in which the audience perceives an authentic social and environmental responsibility on the part of the companies.

It is also a challenge to adopt good practices of effective communication, in which transparent marketing strategies are established and aligned with actions in favor of sustainability, which could generate greater credibility, loyalty, and loyalty in the audience, therefore, social and environmental responsibility campaigns must be authentic and supported by tangible actions to avoid greenwashing, where companies pretend to be more sustainable than they really are.

It is a challenge for brands, the strategy of education and awareness on the part of consumers, since this requires the creation of creative, transparent, relevant, and strategic content that helps consumers understand the importance of the values of sustainability and how their actions contribute to this purpose, this strategy demands from brands a greater understanding and monitoring of the behavior of the audience, promoting communication spaces that allow a better and greater understanding and appropriation of these values and interaction of consumers with the brand.

The transition to more sustainable materials may increase production costs and affect profit margins, however, the opportunity to impact conscious consumers seeking eco- friendly products requires brands to focus on innovation and product design not only as a differentiation strategy but also as a customer loyalty strategy.

Addressing sustainability challenges from the perspective of green marketing strategies involves creating authentic narratives, educating consumers, and adapting to changing market demands, which can result in long-term benefits for companies committed to sustainability.

Conclusion

The objective was to identify and analyze the green marketing strategies implemented by sustainable fashion companies and to identify the impact generated by these strategies on the final consumer.

The results of the research conducted identify how the concept of sustainable fashion has emerged as an alternative to fast fashion, promoting practices that respect the environment, human health, and workers, as well as the use of sustainable materials and local production. In Colombia, the textile sector is an important component of the economy, with steady growth and an increasing focus on sustainability and the SDGs.

The research also shows how consumers show an increasing preference for sustainable fashion due to environmental concerns, highlighting the importance of environmental awareness in purchasing decisions. Sustainable fashion companies must implement effective marketing strategies to communicate their sustainable values and commitment to the SDGs, as well as engage consumers and raise awareness of the importance of sustainability, encouraging active participation in environmental protection.

These factors demonstrate the importance of social networks as communication channels that allow brands to develop their virtual communities around the green marketing strategies designed. Social networks allow consumers to get closer to the brands, learn about the products and the impact generated by the company's processes and the consumption of these products, aspects that are decisive for the end consumer.

The main green marketing strategies applied by these types of companies include educating their consumers on sustainability issues, promoting products made with sustainable materials in their communities, using sustainability seals on their packaging to publicize their relevance, and forming strategic alliances that allow for market growth and brand positioning.

For companies with a business model focused on sustainable fashion, it is essential to have a coherent marketing strategy for the end consumer, avoiding the use of greenwashing practices that do not coincide with authentic green marketing practices, where, in addition to the commercialization of products, the aim is to increase consumer knowledge and awareness of issues related to environmental care and conservation, to bring them closer to the brand's processes, and to ensure that potential customers perceive that the value proposition of the products is aligned with the principles of sustainable development.

Overcoming challenges in sustainability practices in the textile sector involves not only facing obstacles, but also taking advantage of opportunities to differentiate, communicate effectively, educate consumers, innovate in materials and products, and develop green marketing strategies that reflect a genuine commitment to sustainability.

References

Alí García-Capdevilla D, López-De Parra L, Millán-Rojas E (2021) Context and conditions of eco-marketing to promote nature tourism. Revista EIA 1:1–20

Berger-Grabner D (2018) Sustainability in fashion: an oxymoron? Springer, In Innovation Management and Corporate Social Responsibility

Boons F, Lüdeke-Freund F (2013) Business models for sustainable innovation: state-of-the-art and steps towards a research agenda. J Clean Prod 45:9–19. https://doi.org/10.1016/j.jclepro.2012.07.007

Burguillo M, Jurjo J (2007) Ecolabeling: an efficient environmental policy instrument? ICE Economic Bulletin, Spanish Commercial Information, ISSN 0214-8307

Caetano LM (2022) The (un) sustainability of fashion industry: fashion-on-demand as solution? (doctoral dissertation)

Calvo S (2018) !e SDGs: a guide to sustainable fashion. Global Design and Challenges Cartografia

Carey L, Cervellon M-C (2014) Ethical fashion dimensions: pictorial and auditory depictions through three cultural perspectives. J Fash Mark Manag 18(4):483–506. https://doi.org/10.1108/JFMM-11-2012-0067

Castellano S, Urdaneta G, Joheni A (2015) Green marketing strategies used by companies worldwide. TELOS. J Interdiscip Stud Soc Sci UNIVERSIDAD Rafael Belloso Chacín ISSN 1317–0570

Chamorro A (2001) El Marketing Ecológico, [online] 5campus.org, Medio ambiente <http://www.5campus.org/leccion/ecomarketing>

Christen A (2018) To what degree could Swiss millennials' conflicting desire for fashion, brand meaningfulness and sustainability be solved by the fashion-on-demand business model (doctoral dissertation). Haute école de gestion de Genève

Cohen J (2004) Parasocial break-up from favorite television characters: the role of attachment styles and relationship intensity. J Soc Pers Relat 21(2):187–202. https://doi.org/10.1177/0265407504041374

Delmas M, Burbano V (2011) The Drivers of Greenwashing California. ManagRev 54(1):64–87. https://doi.org/10.1525/cmr.2011.54.1.64

Dhir A, Sadiq M, Talwar S, Sakashita M, Kaur P (2021) Why do retail consumers buy green apparel? A knowledge-attitude-behaviour-context perspective. J Retail Consum Serv 59:102398. https://doi.org/10.1016/j.jretconser.2020.102398

Doukas H, Nikas A, González-Eguino M, Arto I, Anger-Kraavi A (2018) From Integrated to Integrative: Delivering on the Paris Agreement. Sustainability 10(7):2299. https://doi.org/10.3390/su10072299

Evans S, Peirson-Smith A (2018) The sustainability word challenge. J Fashion Market Manage: Int J 22(2):252–269. https://doi.org/10.1108/JFMM-10-2017-0103

Grazzini L, Acuti D, Aiello G (2021) Solving the puzzle of sustainable fashion consumption: The role of consumers' implicit attitudes and perceived warmth. J Clean Prod 287:125579. https://doi.org/10.1016/j.jclepro.2020.125579

Herrera A, Campi A, Fariño N (2021) The Harold Lasswell model applied in digital communication. Case: the post

Hwang YY, Jo GY, Oh MJ (2020) The persuasive effect of competence and warmth on clothing sustainable consumption: the moderating role of consumer knowledge and social embeddedness. Sustainability 12(7):2661. https://doi.org/10.3390/su12072661

Kabir SMF, Chakraborty S, Hoque SMA, Mathur K (2019) Sustainability assessment of cotton-based textile wet processing. Clean Technol 1(1):232–246. https://doi.org/10.3390/cleantechnol1010016

Kim J, Kang S, Lee KH (2020) How social capital impacts the purchase intention of sustainable fashion products. J Bus Res 117:596–603. https://doi.org/10.1016/j.jbusres.2018.10.010

Kong HM, Ko E, Chae H, Mattila P (2016) Understanding fashion consumers' attitude and behavioral intention toward sustainable fashion products: focus on sustainable knowledge sources and

knowledge types. J Glob Fash Market 7(2):103–119. https://doi.org/10.1080/20932685.2015.1131435
Kozinets R (2019) Netnography the essential guide to qualitative social media research (THIRD)
Marko M, Kusá A (2023) Greenwashing and the nature of education in relation to consumer trustin fast fashion marketing communication. Https://Communicationtoday.Sk/Greenwashing-and-the-Nature-of-Education-in-Relation-to-Consumer-Trust-in-Fast-Fashion-Marketing-Communication/
Martínez-Barreiro A (2020) Sustainable fashion: beyond scientific bias, a research field of social practices. Society Econ 40. https://doi.org/10.25100/sye.v0i40.7934
Martínez-Rodríguez MC, Mayorga-Pérez O, Vera-Martínez MC, García-Morales MI (2018) Ecolabeling and green products: development and competitiveness. Tecnología En Marcha J 31(2):87. https://doi.org/10.18845/tm.v31i2.3626
Mishra U, Wu J-Z, Sarkar B (2021) Optimum sustainable inventory management with backorder and deterioration under controllable carbon emissions. J Clean Prod 279:123699. https://doi.org/10.1016/j.jclepro.2020.123699
Muñoz-Torres MJ, Fernández-Izquierdo MÁ, Rivera-Lirio JM, Ferrero-Ferrero I, Escrig-Olmedo E (2021) Sustainable supply chain management in a global context: a consistency analysis in the textile industry between environmental management practices at company level and sectoral and global environmental challenges. Environ Dev Sustain 23(3):3883–3916. https://doi.org/10.1007/s10668-020-00748-4
Nielsen J (2019) Micro-level translation of corporate sustainability: When strategy meets practice in the Danish hospitality sector. J Clean Prod
Novillo Díaz LA, Pérez Espinoza MJ, Muñoz JC (2018) Green marketing, trend or fashion? Universidad y Sociedad 10(2):100–105
Obrecht M, Haddad RE, Elbary RA, Lukman RK, Rosi M (2019) Promoting sustainable and circular plastics use in egipt with implementation of ecodesign principles. System Safety: Human—Tech Facility—Environ 1(1):441–448. https://doi.org/10.2478/czoto-2019-0057
Olatubosun SA, Bello S (2024) Time-variant consideration of parameters dependence-based reliability of passive systems: synopsis and proposed framework. Progress Nucl Energy 169:105082. https://doi.org/10.1016/j.pnucene.2024.105082
Pal R, Gander J (2018) Modelling environmental value: an examination of sustainable business models within the fashion industry. J Clean Prod 184:251–263. https://doi.org/10.1016/j.jclepro.2018.02.001
Pal R (2017) Sustainable design and business models in textile and fashion industry. In Sustainability in the textile industry. Springer
Pandit K, Dashti H, Hudak A (2020) Understanding the effect of fire on vegetation composition and gross primary production in a semi-arid shrubland ecosystem using the Ecosystem Demography (EDv22) model. Biogeosci Discuss
Ray S, Nayak L (2023) Marketing sustainable fashion: trends and future directions. Sustainability 15(7):6202. https://doi.org/10.3390/su15076202
Reid E, Duffy K (2018) A netnographic sensibility: developing the netnographic/social listening boundaries. J Mark Manag 34(3–4):263–286. https://doi.org/10.1080/0267257X.2018.1450282
Salamzadeh A, Paramba JN, Brandstrup M, Dana L-P (2023) 19 A sustainable business model for the fashion sector. In fashion and environmental sustainability, pp 237–250. De Gruyter. https://doi.org/10.1515/9783110795431-019
Sánchez-Vázquez P, Gago-Cortés C, Alló-Pazos M (2020) Sustainable fashion and consumer preferences. 3C Empresa. Res Critical Think 9(3):39–57. https://doi.org/10.17993/3cemp.2020.090343.39-57
Sauermann H, Vohland K, Antoniou V, Balázs B, Göbel C, Karatzas K, Mooney P, Perelló J, Ponti M, Samson R, Winter S (2020) Citizen science and sustainability transitions. Res Policy 49(5):103978. https://doi.org/10.1016/j.respol.2020.103978

Schivinski B, Christodoulides G, Dabrowski D (2016) Measuring consumers' engagement with brand-related social-media content. J Advert Res 56(1):64–80. https://doi.org/10.2501/JAR-2016-004

Shrivastava A, Jain G, Kamble SS, Belhadi A (2021) Sustainability through online renting clothing: circular fashion fueled by instagram micro-celebrities. J Clean Prod 278:123772. https://doi.org/10.1016/j.jclepro.2020.123772

Spielmann N (2021) Green is the new white: how virtue motivates green product purchase. J Business Ethics 173(4):759–776. https://doi.org/10.1007/s10551-020-04493-6

Véliz J, Freire RC (2019) El marketing verde the green marketing

Woetzel J, Krishnan M (2020) Climate risk and response. Physical hazards and socioeconomic impacts. McKinsey Global Institute. https://www.preventionweb.net/files/70006_mgiclimateriskandresponsevf.pdf

Xie-Carson L, Magor T, Benckendorff P, Hughes K (2023) All hype or the real deal? Investigating user engagement with virtual influencers in tourism. Tour Manage 99:104779. https://doi.org/10.1016/j.tourman.2023.104779

Chapter 8
Green Marketing and Its Contribution to the Competitiveness of the Textile and Apparel Sector. Case Study in Cali

Magdalida Murgueitio, Eugenio Moreno, Edy Lorena Burbano Vallejo, and Carlos Arturo Vargas Sierra

Introduction

Today's society faces urgent challenges in the face of global warming and climate change, close to the turning point of 1.5 ºC agreed at the Conference of the Parties 28 (COP28) in December 2023, since that year the temperature reached 1 0.48 ºC (National Geographic 2024) as a result of the unbalanced use of finite natural resources. To address the above, there are global frameworks and agreements referred to by the United Nations (2015) in its climate actions, such as the Sustainable Development Goals (SDG) and the United Nations Framework Convention on Climate Change (2015), that require not only political will on the part of the rulers of the nations, but also a society educated in environmental matters. These must provide environmental and economic solutions, in order to "reduce emissions, adapt to climate impacts and finance the necessary adjustments" (United Nations n.d.a, par.

M. Murgueitio (✉)
Human Development, Public Accounting Program at the Universidad de San Buenaventura, Cali, Colombia
e-mail: mmurguei@usbcali.edu.co

M. Murgueitio · E. Moreno · E. L. Burbano Vallejo · C. A. Vargas Sierra
Member of the Research Group On Economics, Management, Territory and Sustainable Development (GEOS), Cali, Colombia
e-mail: emoreno@usbcali.edu.co

C. A. Vargas Sierra
e-mail: cavsierr@usbcali.edu.co

E. Moreno · C. A. Vargas Sierra
Public Accounting Program, Universidad de San Buenaventura, Cali, Colombia

E. L. Burbano Vallejo
Economics Program, Universidad de San Buenaventura, Cali, Colombia

A. E. Toscano-Hernández et al. (eds.), *Sustainability, Stakeholders and Marketing in the Textile Sector*, SDGs and Textiles, https://doi.org/10.1007/978-981-97-6952-0_8

11), where "human industrial activity has caused most of the global warming of the last century" (National Geographic 2024, par. 8.4), and where the textile sector is responsible for 10% of global carbon emissions and 20% of drinking water pollution (Instituto para la Exportación y la Moda [Inexmoda] 2024).

In this regard, the United Nations (2015) raised SDG 6 as one of the global challenges, which states that "water is also a crucial part of adaptation to climate change, and is a decisive link between society and the environment" (United Nations n.d.b, par. 8.1). Various problems converge here as a result of today's modern life governed by consumption, where supply and demand are not balanced and, therefore, affect social, economic and environmental sustainability.

Therefore, faced with multiple problems, this research addressed the textile and clothing industry sector in a case study in Cali, the third city that contributes to the national GDP, after Medellín and Bogotá. This case is associated with green marketing and the competitiveness of three companies that transform fabrics, textiles, fibers and yarns into accessories or clothing.

In accordance with the premise of green marketing, it promotes sustainable products and services, recommends environmentally friendly practices and strategies to satisfy consumer needs and desires, and offers minimal negative impact on the natural environment, because its purpose is to communicate an image of corporate social responsibility with the environment during the marketing of ecological and reusable products (Ministerio de Ambiente y Desarrollo Sostenible 2022; United Nations Organization [UNO] 2022). From the above, green marketing can be seen as a means to contribute to the sustainability of natural resources, as a source of production and competitiveness in different economic sectors.

The textile and clothing industry in Colombia is no exception, because it can generate competitive advantages and responsibly migrate to green companies defined in the National Green Business Plan in force in Colombia. An example of this are organizations that provide goods and services with special characteristics: ecosystem and life cycle approach, positive social and environmental impacts, sustainable practices, GHG reduction, etc. All of the above, "from the use, transformation, valorization and conservation of resources to contribute to the development of the territories" (Ministerio de Ambiente y Desarrollo Sostenible 2022, p. 94).

Thus, green marketing, seen as a process to identify, anticipate and satisfy consumer needs, has the challenge of meeting indicators in profitability, responsibility and sustainability; and this is a bet inherent to all economic sectors. Thus, for this research, and in accordance with the evolution of green marketing, the objective is to explore how green marketing, with its practices and strategies, allows companies in the textile and apparel sector to generate competitive advantages and migrate responsibly according to the proposal of the National Green Business Plan in force in Colombia (Ministry of Environment and Sustainable Development, 2022).

Textile and Clothing Sector in Colombia

> The textile industry in Colombia is one of the most representative and traditional. It has had a history of more than 100 years, in which some aspects have stood out: the creation of a diversity of representative companies at an international level, the government plans for the sector, the penetration in different areas such as the academy, inclusion in productive transformation programs, and creation of a world-class sector as shown in Fig. 8.1 (Sectorial, 2022).

Likewise, Procolombia (2022), as cited in Sectorial (2023) reported that the textile and clothing sector represents 9.4% of industrial GDP and provides employment to more than 600 000 people. On the other hand, there are 6195 of these companies mostly concentrated in Medellín, Bogotá and Cali. This sector is divided into the areas of clothing manufacturing, preparation, dyeing and product sales. According to the report on the Colombian textile sector, an economic reactivation has been reported by exceeding the figures before the pandemic: the contribution to GDP in 2022 grew by 17.6% compared to 2021; regarding the purchase of clothing, this increased by 9.4% compared to 2022; textile exports increased by 11.7%; in clothing, there was an increase of 5.0%; and imports increased with 20% and 29% in textiles and clothing respectively (EMIS 2023).

According to Fig. 8.1, Bill 218 of 2022 (Proyecto de Ley 218 de 2022) is being processed in the plenary session of the House of Representatives, the purpose of which is:

> Establish a comprehensive textile waste management system for large textile-producing or marketing companies, in order to promote the circular economy, mitigate climate change, reduce the water footprint and relieve the pressure on landfills. (Cámara de Representantes de Colombia 2022, par. 8.1).

This initiative is necessary to regulate and safeguard Colombia's precious biodiversity, since the textile and clothing sector is the one with the greatest negative impact on the territory's environment; hence, the need to promote and migrate towards a circular economy in the processes and behaviors in its supply chain. Added to the

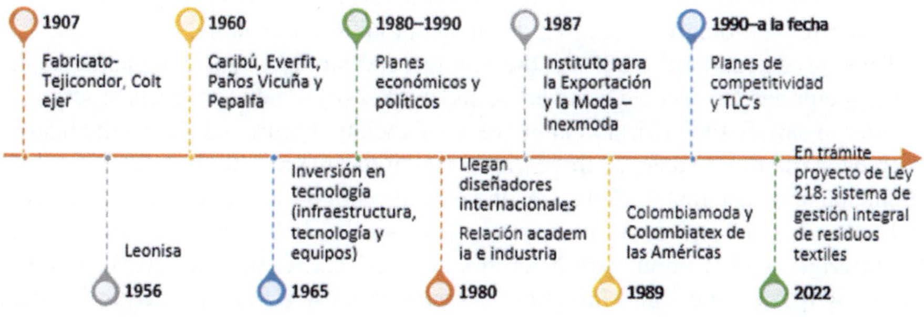

Fig. 8.1 History of the textile industry in Colombia. *Note.* Adapted from *Historia del negocio textil en Colombia*, by Sectorial 2022

above is communication, with a more sensitive consumer willing to modify consumption patterns, which is associated with "the gap between intention and behavior towards the purchase of ecological clothing" (Rasheed et al. 2024, p. 1); that is where green marketing plays a relevant role in improving the brand image of companies that are committed and responsible to the environment.

In this process, ISO 14006/2011 Standard (International Organization for Standardization [ISO] 2011) "provides guidelines to help organizations establish, document, implement, maintain and continually improve their eco-design management as part of an environmental management system (EMS)" (p. 10). This is applicable to the value chain: from sowing to the final consumer, because sustainability is an imperative need for future generations (Ruppel et al., 2021). In line with the above, at Colombiatex 2024, "sustainability and circularity emerged as crucial themes, reflecting the preparation of the national industry for an imminent transformation" (Inexmoda 2024, p. 27); and this is scalable to organizations, regardless of their size or activity, as in homes in general.

Green Marketing

Brands today face a consumer who is sensitive and aware of their environmental environment and their quality of life, which demands a preference for products and services that respect and preserve the sustainability of the environment. For this reason, brands have sought to stay in the market; and, consequently, they have opened new projects by starting on the path towards sustainable companies, with the support of the possibilities offered by green marketing as the means to communicate and inform sustainable actions of business activities according to social evolution.

Historically, textiles and fashion have consumed important natural resources, and this figure is proportional to the negative impact on the environmental sustainability of the planet, due to excessive offers and demands in the sector. Consequently, the sustainability of the textile sector goes beyond achieving the green label: it is about environmental, social and economic transformation; with sensitive and responsible consumers towards the preservation of natural resources (Atria Innovation, 2023; Comunicare Neuromarketers, 2022; Pacto Mundial Red Española, n.d.).

Thus, green marketing defines the commercial strategy that companies in the textile sector need to execute to advertise and market sustainable products or services, in order to satisfy the environmental needs of society. Figure 8.2 shows the history of green marketing to date, so three transformations are presented: ecological, environmental and sustainable. This way, the execution of strategies must be consistent with what is said and done; thus, strategies and decisions must be aligned with the messages sent (Comunicare Neuromarketers, 2022; Ecocreare, 2023) because, according to the American Marketing Association (AMA, n.d., as cited in Creative Corner, 2021), "[green marketing has been] defined as follows: 'the development and marketing of products designed to minimize negative effects on the environment, or to improve their quality'" (par. 8.2).

Fig. 8.2 History of green marketing and its relationship with sustainability. *Note* Adapted from Comunicare Neuromarketers 2022; Ecocreare 2023; and Kumar et al. 2017

Green marketing promises economic and market advantages for companies, employees, consumers and the environment; this, through the communication of strategies that involve the three dimensions of sustainability: economic, social and environmental. Likewise, it seeks to integrate those involved in the supply and demand of the textile and clothing sector into the responsible culture. Therefore, marketing evolves according to changes in trends and assumes "a new concept that integrates the needs of the company, people and society" (Uribe 2023, p. 291), without implying to solve the needs of future generations. Consequently, companies and consumers have understood that all human beings are socially responsible and part of the problem (Robledo et al. 2021), which is why the collaborative work of actors in the textile and clothing system is required, in order to guarantee permanence in the market and contribute to the sustainability of the planet and the well-being of the territories.

Continuing with this line, it is worth remembering that the Colombian textile sector in the era of green marketing has a sensitive consumer in the textile value chain, which is why they participate especially in three stages: the first is the decision to purchase garments with low or high-costs, quality and duration; the second, the care of the product in consumption referring to washing, drying and ironing; and the third, that of responsible disposal habits after consumption. The above highlights the benefits of purchasing slow fashion products: sustainable forms of product care and virtuous reuse and recycling practices (Ahmed 2020). Still, according to this author, women tend to change fashions quickly, so some clothes have a shorter lifespan than others.

Sustainability

Talking about sustainability implies a balanced management between environmental, social and economic aspects in any sector of the economy, and the textile and clothing sector has a high commitment to the environment, because that is where the resources for the business operation are obtained from. These, managed with conscious responsibility and in balance with nature, lead to sustainable development and the coordination of all actors in the textile sector, in line with a consumer increasingly sensitive to the environment.

According to Sadiq et al. (2023), by increasing environmental sustainability, organizational performance increases, which helps expand industrial competitive advantage; and, this way, the integration of sustainability and green marketing in the sector is associated with different approaches adopted by companies. This is the case of the concept of circular economy, especially because this sector is considered one of the biggest polluters for the environment.

In this order of ideas, the sector needs to innovate in processes and improve nature-friendly practices in its production and marketing chain, especially with some SDGs, with the aim of lowering the negative impact associated with the use of water, energy, decent work, global warming and climate change:

- SDG 6. Clean water and sanitation. The textile sector is responsible for 20% of global wastewater, hence the relevance of the EMS and the adoption of Bill 218 of 2022 for the creation of the comprehensive textile waste management system for large textile producing or marketing companies. Thus, although small and medium-sized enterprises (SMEs) are not required, this would be an excellent adoption at the scale of their possibilities.
- SDG 7. Affordable and non-polluting energy. The call to the sector implies adopting clean technologies.
- SDG 8. Decent work and economic growth. It implies guaranteeing legal conditions in the territories to promote well-being and avoid inequalities and damage to the environment.
- SDG12. Responsible production and consumption. The aim is to change the production and consumption model to reduce and avoid the waste of natural resources.
- SDG 13. Climate action. It is imperative that all social actors take measures; otherwise, there could be no social and economic progress (United Nations 2015).

This way, the sustainability of companies is the language to ensure their future (Pacto Mundial Red Española, n.d.); and, to plan it, the environmental impact must be reduced, the working conditions of employees must be improved, ethical criteria must be applied in the supply chain and the design of sustainable products beyond green labels and movement towards sustainability in the industrial sector of textiles and clothing in Colombia. All this, to reduce environmental impact, improve workers' conditions, and adopt ethical practices in the supply chain, sustainable design and production (Atria Innovation 2023).

Given this, it should be highlighted that sustainability is a process where factors and actors intervene that, in the dynamics of supply and demand, define the future of natural resources and the well-being of the population, since there is a need to prioritize quality above quantity when it comes to the consumption of sustainable clothing to protect the next generations. Likewise, you can contribute to business sustainability in economic-financial terms by creating social and environmental value. This is why the incorporation of regulations over comprehensive textile waste management systems when producing and marketing companies is so important, and this was proposed in Bill 218 of 2022. Added to the above, there is the need to make way for the transition from traditional to green businesses, as well as promoting new green initiatives.

The change in the environment-business interaction today focuses on a dynamic interrelation towards sustainability to protect natural resources in the territories and contribute to their local development. Therefore, business sustainability requires taking measures and active participation in sectoral, intersectoral and international cooperation initiatives; hence, the position that true sustainability can only be achieved if the interdependent needs of all species of current and future generations are met, and this could be called "multispecies sustainability" (Rupprecht et al. 2020). The above implies recognizing that nature is finite and human beings depend on other living beings (human or not) to balance responsible consumption in satisfying their needs.

Method

The objective of exploring how green marketing, with its practices and strategies, allows companies in the textile sector to generate competitive advantages and responsibly migrate to the sector in question in accordance with the National Green Business Plan in force in Colombia (Ministerio de Ambiente y Desarrollo Sostenible 2022). Based on a phased design in SMEs registered and actively working in the city of Cali. In this regard, it is worth noting that the department of Valle del Cauca to date has a representation of 8.7% in the sector, with 1 854 086 active companies, according to data from the first 2 months of 2023 (Ministerio de Comercio, Industria y Turismo [MINCIT], 2023).

> Decree 957 of 2019 established the criteria for classifying companies as micro, small or medium-sized companies, according to the income received in a year from ordinary activities and the sector in which they are located.
>
> firms that do not classify in any of the previous three sectors must use the thresholds determined for the manufacturing macro-sector.
>
> In the case of companies that have more than one economic activity, the size will be defined by the one that reports the highest volume of income. (Actualícese, 2023, par. 8.1, 8.12–8.13).

- Phase 1: search and selection of bibliographic material.
- Phase 2: selection of three SMEs from the textile and clothing sector taking into account two conditions: because of their size (micro, small and medium) and because in Colombia:

> Between January and March of 2023, micro-businesses represented 95.3% of the country's business fabric, and 3.5% of small businesses and medium and large businesses accounted for 0.9% and 0.3% of the national total, respectively. (MINCIT, 2023, par. 8.1).

- Phase 3: application of semi-structured interviews to three entrepreneurs in the textile and clothing sector.
- Phase 4: Contrasting bibliographic documents, marketing campaigns and semi-structured interviews.

Results

The businessmen were questioned about the marketing strategies and techniques implemented in their businesses based on the seven suggested by green marketing (APD, 2021) and the five competitive factors of the textile and clothing sector in Colombia (Procolombia 2022) analyzed from different disciplinary perspectives of the research team, such as those of the ISO 14006/2011 Standard (ISO 2011), which promotes eco-design as a philosophy to create sustainable and environmentally friendly products and services:

> Eco-design can be understood as an integrated process within design and development, which aims to reduce environmental impacts and continuously improve the environmental performance of products, throughout their life cycle, from the extraction of raw materials to the end of its useful life. To benefit the organization and to ensure that the organization achieves its environmental objectives, eco-design should be carried out as an integral part of the organization's business operations. Eco-design can have implications for all functions of an organization.
>
> [...]

The general areas of knowledge required to incorporate eco-design within an EMS are:

(a) *The evaluation of the impacts of products on the environment.*
(b) *The identification of appropriate eco-design measures to reduce the adverse effects of these environmental impacts.*
(c) *The design and development process and how the eco-design process and its management are integrated into an EMS. (p. 8)*

The case study showed that medium-sized companies have taken the first steps to adopt ecodesign in their EMS, while small companies see it as a long-term opportunity and micro companies do not have it in their improvement processes, but recognize the importance of it in their interest to remain in the sector.

Regarding the green marketing strategies applied in companies, these point to various developments in each of them, based on the marketing campaigns implemented to date.

Strategy 1: communicate what is being done and why.

The advertising campaigns fully comply with the communication about what they do and why, in order to inform clients about new trends in the business and the sector that is progressively moving towards a sustainable and nature-friendly model. These topics have strengthened the brand and image of the company in the niches in which these operate. Strategies 1 and 2 converge with SDG12 in informing and educating the customer on the use of garments and accessories to prolong their useful life, contributing to the peremptory campaign to be friendly to nature by guiding responsible consumption to reduce and diminish waste.

Strategy 2: promote responsible consumption.

The promotion of responsible consumption is materialized with recommendations for washing and treating garments to extend their use. This is a standard requirement demanded in marketing; but, according to changing customer needs, companies have been under pressure to incorporate elements to promote responsibility in their use.

Strategy 3: get and display certifications and labels.

The certifications and labels meet the requirements demanded by legal regulations in the operation of the business; this, depending on its size, goals and market niche. Likewise, these recognize that the new guidelines contained in Bill 218 of 2022 must be strengthened, aimed at the comprehensive management of textile waste, which promotes the circular economy to achieve environmental and business sustainability. In this regard, contributions must be reviewed and adjusted to "mitigate climate change, reduce the water footprint and relieve the pressure on landfills" (par. 8.1).

Strategy 4: participate in events and causes in favor of the environment.

In this strategy, entrepreneurs agree on their low participation in environmental events, but recognize the adoption of internal activities to improve the environmental topic. Even so, these are not communicated to the client, even though they are a dissemination strategy that could help the image of the business.

Strategy 5: spread the company values.

Regarding strategy 4, entrepreneurs are linked to environmental events and causes associated with the company's values, because such link provides a positioning in the market with a client that is increasingly sensitive to sustainability. For the case study, the medium and small companies with 15 and 10 years of existence in the market, respectively, incorporate information on their web page associated with their values and environment. The micro company does not have this communication channel, but does have social networks.

Strategy 6: create green marketing business alliances.

At this point, the formation of business alliances is incipient in terms of working towards business sustainability, given the wide gaps between the companies studied associated with income and maturity in the execution of sustainable strategies.

Strategy 7: use sustainable packaging.

The packaging used to date is not biodegradable, which is why it continues to contribute to environmental pollution. In accordance with the previous strategies, the three businessmen interviewed from the textile and clothing sector agreed that, although they are aware that their production and marketing processes are environmentally friendly, these involve high operating costs to remain in the market. Likewise, the total adoption of the seven green marketing strategies is a medium-term perspective based on the competitive possibilities it can bring to the organization to achieve sustainability in business (APD, 2023).

Regarding the ISO 14006/2011 Standard (ISO, 2011), it specifies certain differences in processes for organizations according to their size, which determines their financial capacity to operate and invest in new nature-friendly trends, which favors the economy of the organization. The above arises from the incorporation of eco-design and involves three general areas of knowledge: evaluating the impacts of the product or service on the environment; identifying eco-design measures to reduce the negative effects of environmental impact; and the process of development and integration of eco-design and its management in an EMS.

Other topics addressed with the businessmen were the five competitive factors of the textile and clothing sector in Colombia (Procolombia, 2023), directly related to business projection, the investment capacity of these companies, and market supply and demand. Here the relationship was established between the contributions of green marketing strategies and behavior to competitiveness, which would facilitate the transition to environmental and business sustainability. Also, the businessmen referred to the various developments according to various factors, which are shown below.

Factor 1: innovation and diversification.

Innovation and diversification in raw materials are directly related to imports, which represent a high cost in business operations. Such factor limits the level of innovation and diversification to keep up with the changing demands of the market and those of a more environment-sensitive client.

Factor 2: experience and reliability.

Experience and reliability, goals that every company needs to achieve and those chosen for this study are no exception. These refer to continuous work, due to the changing dynamics of the market and the greater demand in terms of quality and price on the part of the consumer; this also helps to the rise of the power of social networks to position brands and the image of companies. The challenge is great, especially when the sector has a long history and consolidation of designers, brands, clothing companies, yarns and textile producers.

Factor 3: strategic location and commercial advantage.

The comparative and competitive advantage of the country as a competitive platform, according to Procolombia (2023), provides the sector with facilities for its commercial operation. This, for the companies under study, offers various possibilities to trade and compete in similar market niches. However, the marketing strategies employed are no guarantee of achieving a commercial advantage over competitors in the constant battle for prices and discounts to attract more customers.

Factor 4: competitive prices and quality raw materials.

The advantages of factors 2, 3 and 4 are associated with competitiveness and determine high-quality products and services for the consumer, which in turn allows the latter to achieve reliability and loyalty. The above contributes to companies' permanence in the market and to the dynamization of the economy, which, for the companies in the study, are changing due to the externalities that they must face daily.

Factor 5: commitment to sustainability.

The sector recently adopted measures for the transition towards environmental sustainability in harmony with the corporate purpose of the company, which is why today it is imperative to review, adjust and execute strategies that are friendly to the use of natural resources, in order to guarantee the well-being of future generations. The above must be a gradual process, which requires the collaborative work of all those involved in the sector's value chain. For the case study, the three companies are aware that the route to remain in the market is sustainability and for this, they initiated processes with different degrees of development according to the incorporation of the EMS guidelines and SDGs 6, 7, 8, 12 and 13.

Analysis and Discussion

The comparison of green marketing strategies and competitive factors in the three companies studied showed that in order to comply with factors 4: competitive prices and quality raw materials and 5: commitment to sustainability, it is necessary to adopt strategies 6: create green marketing business alliances, 7: use sustainable packaging and 2: promote responsible consumption, to achieve the expected results in competitiveness based on the contributions of green marketing in the expansion and customer loyalty.

In this order of ideas, green marketing as a strategy has evolved: it no longer seeks only to increase the number of clients, but also aims to communicate good environmental practices in companies, which contributes to the positioning of the brand and corporate image. With this, new products have been consolidated, these are second-hand ones; this market is increasingly gaining strength not only due to low costs, but also due to environmental, social and economic sustainability. When crossing the contributions of green marketing with the competitiveness of the companies studied as a sample, certain coincidences and gaps are observed according to their size, the challenges when overcoming the externalities of the economy that affect the decisions and projection of the organizations, and the constant change in trends and consumer demands in their cost-quality relationship, which determines positioning and permanence in the market.

In that sense, the process of adopting green marketing as a tool to generate competitive advantages and responsibly migrate to green companies is a path that requires strengthening the cultural, technological and managerial aspects, with the support of government authorities. Added to the above are the purposes of some of the SDGs, such as the eighth, aimed at inclusive and sustainable economic growth, in order to generate green jobs and decent work. Likewise, it is essential to move towards the green economy, due to the emergence of green professions, and the traditional profession must be motivated so that it also becomes green.

Now, the business and environmental future depends on the degree of commitment and joint work of all actors to guarantee the sustainability of the planet; and the textile sector, as the largest polluter, must continue with its transformation and commitment to the green route, hand in hand with marketing and with the addition of a greater number of green consumers. These are the ones called to raise the flag for the execution of plans to reduce environmental impact, by pressuring companies to assume coherence between what is said and what is done in their green marketing campaigns in favor of environmental sustainability. In short, organizations must adopt ethical practices in the supply chain, design, production and effective communication with the green consumer.

Consequently, marketing has become a business resource to achieve profitability, configure an image and an ethical brand that is responsible for the environment, and offer the green consumer a product or service with convenient quality and price. Compliance with the above would lead to the strengthening of competitive advantages for the organization, as well as a greater permanence in the market, the generation of jobs, and the contribution to territorial development.

Conclusions

The case study in Cali on green marketing and its contribution to the competitiveness of the textile and apparel sector showed that it requires awareness on the part of entrepreneurs about environmental responsibility in the use of natural resources in the daily operation of the business and requires investing in the convergence of green marketing strategies and competitive factors to travel the green road.

Caring for the planet first requires a citizen with environmental education, willing to responsibly use nature's resources for personal and commercial benefit. On the other hand, State policies and conditions are needed, so that the community and entrepreneurs can adopt the necessary measures in order to move towards green organizations and contribute to the development of the territory.

The contribution of green marketing to competitiveness in the case study in Cali is directly related to the ability to manage a sustainable company with the possibility of increasing revenues and customers. Going green in the companies studied is an act of faith, common sense and respect for preserving the source of resources that allow the daily operation of the business.

References

Actualícese (2023) Así operará la clasificación de empresas por tamaño en 2023. https://actualicese.com/asi-operara-la-clasificacion-de-empresas-por-tamano-en-2023/

Ahmed M (2020) Consumer engagement and roles for sustainable textiles. In: Matthes A, Beyer K, Cebulla H, Arnold M, Schumann A, Sustainable textile and fashion value chains: drivers, concepts, theories and solutions, pp 237–253. Springer International Publishing

APD (2023) Estrategias de marketing sostenible para empresas: conoce las más importantes. https://www.apd.es/estrategias-de-marketing-verde/

Atria Innovation (2023) Sostenibilidad en el sector textil. https://www.atriainnovation.com/sostenibilidad-en-el-sector-textil/

Cámara de Representantes de Colombia (2022) Proyecto de Ley 218 del 29 de septiembre de 2022. Gaceta N° 1190 de 29 de septiembre de 2022. [Por medio de la cual se crea el sistema de gestión integral de residuos textiles para grandes empresas productoras o comercializadoras de textil]. Bogotá

Comunicare Neuromarketers (2022) Historia del marketing verde. https://www.comunicare.es/historia-del-marketing-verde/

Creative Corner (2021) La estrategia de marketing verde. https://creativecorneragency.com/es/la-estrategia-de-marketing-verde/

Ecocreare (2023) Historia del marketing verde y sostenible. https://www.ecocreare.com/historia-del-marketing-verde-y-sostenible/

EMIS (2023) Colombia—informe sector textil y confecciones. https://recursosdigitales.usb.edu.co:3585/php/search/docpdf?pc=CO&sv=EMIS&doc_id=778344190

Instituto para la Exportación y la Moda [Inexmoda] (2024) Observatorio de moda 2024. Inexmoda: https://www.inexmoda.org.co/wp-content/uploads/2024/02/Observatorio-Inexmoda-Febrero-2024-F.pdf

International Organization for Standardization [ISO] (2011) ISO 14006–2011 Standard. ISO

Kumar V, Lahiri A, Dogan O (2017) A strategic framework for a profitable business model in the sharing economy. Ind Mark Manage 69(1):147–160. https://doi.org/10.1016/j.indmarman.2017.08.021

Ministerio de Comercio, Industria y Turismo [MINCIT] (2023) Las microempresas fortalecen el tejido empresarial colombiano. https://www.mincit.gov.co/prensa/foto-noticias/microempresas-fortalecen-el-tejido-empresarial

Ministerio de Ambiente y Desarrollo Sostenible (2022) Plan Nacional de Negocios Verdes 2022–2030. Banco Interamericano de Desarrollo

National Geographic (2024) ¿Qué es el calentamiento global? https://www.nationalgeographic.es/medio-ambiente/que-es-el-calentamiento-global

Pacto Mundial Red Española (n.d.) Sostenibilidad empresarial. https://www.pactomundial.org/que-puedes-hacer-tu/sostenibilidad-empresarial/

Procolombia (2022) Moda del futuro: tendencias y comportamientos de los consumidores. https://prensa.procolombia.co/colombia-hoy/moda-del-futuro-tendencias-y-comportamientos-de-los-consumidores

Rasheed N, Sabir R, Mahmood H, Rauf A, Ibrahim A, Naseem W (2024) Impact of pro-environmental values on sustainable green apparel buying behavior in Pakistan. Cleaner Responsible Consumption 12(1):1–10. https://doi.org/10.1016/j.clrc.2024.100180

Robledo J, Loaiza L, Giraldo N, Reyes T, Rojas A (2021) Marketing verde como estrategia para promoción de la responsabilidad social y ambiental en consumidores del sector alimentos empacados. In: Escobar J, Gómez L, Marketing verde, responsabilidad social y composición empresarial, pp. 13–63. Editorial Universidad Santiago de Cali

Ruppel C, Canetti R, Monacchi C (2021) Adapting sustainable design assessment tools for local development: some insights into Argentina's textile and clothing industry. In: Gardetti M, Larios-Francia R, Sustainable fashion and textiles in Latin America (pags. 25–48). Springer

Rupprecht C, Vervoort J, Berthelsen C, Mangnus A, Osborne N, Thompson K, Urushima A, Kóvskaya M, Spiegelberg M, Cristiano S, Springett J, Marschütz B, Flies E, McGreevy S, Droz L, Breed M, Gan J, Shinkai R, Kawai A (2020) Multispecies sustainability. Cambridge University Press. https://doi.org/10.1017/SUS.2020.28

Sadiq N, Pantamee A, Khudoykulov K, Thi Ngan T, Tan L (2023) El papel de la gobernanza social y ambiental en el logro de los objetivos de desarrollo sostenible: evidencia de los países de la ASEAN. Investigación Económica-Ekonomska Istraživanja, 36(1):170–190. https://investiga.banrep.gov.co/es/content/economic-research-ekonomska-istrazivanja

Sectorial (2022) Historia del negocio textil en Colombia. https://www.sectorial.co/articulos-especiales/item/50352-historia-del-negocio-textil-en-colombia

Sectorial (2023) El sector textil en Colombia ocupa 600 000 personas. LinkedIn: https://es.linkedin.com/pulse/el-sector-textil-en-colombia-ocupa-600000-personas-sectorial-s-a-s

United Nations General Secretariat (2015) United Nations Framework Convention on Climate Change (Paris Agreement). United Nations

United Nations Organization [UNO] (2022) The Johannesburg summit: development on the defensive. https://unctad.org/es/Docs/aconf199d20_sp.pdf

United Nations (2015) Sustainable Development Goals. https://www.un.org/sustainabledevelopment/es/objetivos-de-desarrollo-sostenible/

United Nations (n.d.a) What is climate change? https://www.un.org/es/climatechange/what-is-climate-change

United Nations (n.d.b) Globales issues: water. https://www.un.org/es/global-issues/water

Uribe A (2023) La responsabilidad del marketing verde. Tendencias, 24(2):288–306. https://revistas.udenar.edu.co/index.php/rtend/article/view/8168/9463

Chapter 9
The Green Swan's Theory and Circular Economy and It's Influence in Regenerative Capitalism

Carlos Franciso Ortiz-Paniagua, Norma Laura Godinez Reyes, and Enrique Esquivel-Fernández

Introduction

In its primary definition, climate change (CC) is: "A change of climate attributed directly or indirectly to human activity that alters the composition of the global atmosphere and that is in addition to natural climate variability observed over comparable time periods" (CMNUCC 1997).

According to this definition, it can be said that climate is a variation of the earth's climate with respect to the climatic history that can occur on different time scales and on climatic parameters such as pressure, temperature, relative humidity, winds, precipitation, cloudiness, and all these alterations can be attributed to natural causes and anthropogenic action. In other words, the modification of the climate in a given place with respect to its climatic history are changes of a natural order, however, they are currently associated with human impact on the planet (IPCC 2023).

CC is not a synonym for global warming since it responds to diverse causes and results in multiple consequences, commonly we find it associated with this phenomenon of an increase in the average temperature in the atmosphere and in the oceans. But in addition to warming, it also influences precipitation, cloudiness and many other natural phenomena that directly and indirectly affect the stability of natural ecosystems.

There are several theories about CC, however, there are two theories that have become very important and are contrary to each other. On one hand, there are skeptics who do not believe that there is a change in the climate but choose to say that CC is a

C. F. Ortiz-Paniagua (✉) · N. L. G. Reyes
Universidad Michoacana de San Nicolás de Hidalgo, Mexico, Mexico
e-mail: carlos.ortiz@umich.mx

E. Esquivel-Fernández
Centro de Estudios Sociales y de Opinión Pública de la H. Cámara de Diputados, México, Mexico

A. E. Toscano-Hernández et al. (eds.), *Sustainability, Stakeholders and Marketing in the Textile Sector*, SDGs and Textiles, https://doi.org/10.1007/978-981-97-6952-0_9

process alien to man and that it is a normal behavior of the planet. The other theory explains how anthropogenic activity has significantly affected the planet's normal cycle, altering the climate and precipitating the climatic phenomena that today affect humanity and its quality of life.

The theory of CC due to anthropogenic activity is the one that tries to explain the increase in temperature evidenced in the past 200 years through the emission of large amounts of carbon dioxide into the atmosphere, due to human activities. According to the theory, an essential element in the regulation of the planet's temperature is the greenhouse effect. This effect consists of the existence in the atmosphere of gases, among them carbon dioxide, capable of maintaining in the interior of the planet a part of the solar radiation that reaches the earth's surface and returns in the form of infrared radiation. The concentration of these gases, specifically carbon dioxide, has increased considerably because it is a gas that is emitted in a large number of industrial processes as a consequence of the occurrence of combustion processes (Lopez 2007).

Undoubtedly, the upward trend in the volume of world production, from the beginning of the capitalist system to the present day, is directly proportional to the growth trend of environmental pollution, while there was no mass production, the ecology was in balance, as capitalism was consolidating the destruction of nature was increasing.

As a result of this series of changes in the climate due to human activity, the term sustainability arises. In order to know how the sustainable development approach begins and evolves, it is important to mention the origin of environmental pollution, which is rooted in the Industrial Revolution, since it not only unleashed the scientific and technical economic boom, but with the beginning of it, the intensive, extensive and irrational use of natural resources in search of models of accelerated economic growth was promulgated (Cordero 2012).

Sustainability today is a term linked to the action of man in relation to his environment. Within the ecological discipline, sustainability refers to biological systems that can conserve their diversity and productivity over time. Sustainable development considers not only the social, political and economic conditions of the social group, but also incorporates the human vision of a development in favor of the care of the environment and the natural surroundings in which human beings live (Stern 2016).

Literature Review

In general terms, the CE model consists of replacing a "linear economy", based on the extraction of natural resources and raw materials, the production of goods and services, consumption and the generation of waste or residues, which ultimately generates a negative impact on the environment; this approach proposes both the conversion of waste into new resources, as well as an innovative change in the production system, whereby the design of each phase of the production process is guided by the idea of regeneration, aiming to close loops in industrial ecosystems and minimize waste generation (Cerdá et al. 2016; Chaves Ávila 2018).

When analyzing the CE, then, there is no doubt that its objective is to achieve sustainable well-being in an economy that will last over time; however, to achieve this, it is first essential to study the laws and patterns that the cosmos uses to build the healthy networks that exist among us. If the global economy is to be sustainable, it must not be limited to measuring desired productivity. It is imperative to learn to identify the underlying factors that generate systemic health over time. So then, the economy and the financial system on which it will be based must be designed to incorporate those principles.

Today, the prevailing so-called "free market" economic system is built around a narrow interpretation of Adam Smith's "invisible hand" metaphor, assuming that the actions of individuals and firms in a market economy are miraculously aimed at providing an optimal allocation of resources and thus, achieving the highest level of welfare for society; these unquestioned beliefs include that the economy is separate from the biosphere and the environment, and that maximizing profits for shareholders, GDP growth, as well as optimizing material benefit for consumers, lead to prosperity (Coeure 2020).

People view of the economy separately from other parts of society and the biosphere often ignore or separately address the damage done to other parts of society and the biosphere. Most leading business schools, for example, still teach the mechanistic idea that optimizing short-term shareholder value should be the fundamental goal of a company. This idea carries a clear danger to the health of human communities and life on Earth because it assumes that business is separate from society as a whole, as well as from the biosphere that performs life-enabling functions on which business, employees and customers depend (Samuelson 2006).

Regenerative Capitalism does not seek to end capitalism, but rather to evolve it to produce prosperity and vitality in an inclusive and broadly shared way, just as its founders envisioned it. Importantly, the laws of systemic health that are being discovered can help this new generation of capitalism produce durable and vibrant human networks upon which economic vitality depends (Sahakian 2016).

Methodology

This research uses a multilevel qualitative conceptual descriptive analysis of contrasting models regarding CC, with a first level based on concepts related to economic externalities and sustainability, with an emphasis on the so-called green swans; second level discusses the disruptive economic effects; finally, a third level engages in a discussion about overall risks and the potential mitigation using an approach focused on circular models, with information that comes from previous studies carried out by international organizations and top tier cited papers, as it is shown in the following Fig. 9.1.

I	II	III
Climate change as a market failure	Disruptive economic effects of CC	Overall risks and potential mitigation
First level (Iberdrolla, 2020; Coeure, 2020; Terceiro, 2018; Taleb, 2007; Singhania, Gupta, Chadha, Braune, Dana, & Idowu, 2023; Elliott, 2023; Guo, Li, Wu, & Hao, 2023; Bolton, Després, da Silva, Samama, & Svartzman, 2020).	**Second level** (Svartzman, Bolton, Despres, Pereira Da Silva, & Samama, 2021; Ghosh, Kundu, & Dilip, 2021; Evans, 2014; Posthuma, et al 2024)	**Third level** (Carré, 2017;Cerdá & Khalilova, 2016; Ellen MacArthur Foundation, 2015; Bolton, 2020, Reed, 2009; Bingler & Colesanti, 2022)

Fig. 9.1 Three-level descriptive analysis for CC in the literature. *Source* Own elaboration (2024)

The table above shows the collection of selected papers that are the basis for the critical analysis in relation to CC and green swans, on three levels, from the market perspective, the disruptive effects of this rising phenomenon, how such an approach can help address environmental risks and the potential for mitigation for such issues.

Results

First Level: Climate Change as a Market Failure

In the literature, a market failure is considered a situation that occurs when the market is unable to allocate resources efficiently (Iberdrolla 2020; Coeure 2020). According to Terceiro Lomba (2018), climate change represents an evident market failure, considering it as a global public good with specific properties, including non-rivalry (since the enjoyment of one does not prevent the enjoyment of another) and non-exclusion (since it is freely accessible to all individuals); under this premise, the author considers that climate change is shown as a negative externality of the market, since society as an economic entity, by not considering climate as a common good, has provoked high CO_2 emissions, which as a consequence causes global warming. These actors have not considered the inclusion in the pricing of their products and/

or services of the costs they incur by deteriorating the environment, and inflicting damages to other economic subjects.

Some of the negative externalities caused by this market failure include greenhouse gas (GHG) emissions reaching the atmosphere and remaining there for long periods, conditioning the future climate, uncertainty in dealing efficiently with the problem of climate change, as well as impact and its potential capacity to cause radical and irreversible changes to the environment.

As discussed previously, the concept of "green swan" comes from the concept of "black swan" coined by Nassim Nicholas Taleb in 2007 (based on the past belief that swans all over the world were white, and later with the appearance in Australia of black swans); this author argues that derived from the common global consciousness or in other words the way in which the human being reasons, referring specifically to reasoning in economic terms, this is carried out in an inductive way, meaning that, it is assumed that past information will reveal with certainty or at least very close to the reality of future events, and for this reason, it is essential to introduce the concept of skepticism or distrust in the inductive reasoning used when making a prediction of future events.

Such metaphor shows that, although induction can provide us with conclusions that are probable, at the same time easy and predictable, from a realistic perspective the world does not behave that way, since while white swans are predictable and probable, black swans, (or unexpected events that alter the system) should never be discarded (Taleb 2007).

However, if this concept is transferred to the financial, economic, or social events, we can begin to contemplate various ramifications to what humanity knows as normality; in early 2020, the Bank for International Settlements associated such a concept with the occurrence of unexpected and rare events caused by global warming that implied high financial costs for the global economy and it is called the green swan.[1]

Since climate change and climate finance are strongly attracting attention from scholars and practitioners, subjects such as carbon neutrality, accounting for sustainability, planetary boundaries framework, sustainable finance, managing climate risk for the third pole, financial innovation and green finance, green swans, represent emerging domains of climate finance subjects in the field (Singhania et al. 2023); although the recognition of this situation from some central banks is being part of the current trends, institutional investors, asset managers and banks persist with a conceptualization of climate change as a risk that can be measured, managed, and hedged against, including in their climate policy proposals (Elliott 2023).

For instance, in the case of China, the adoption of a market-oriented economy to encourage green development is following a carbon emissions trading scheme as a strategy that serves as a response to the potential green swans event in the future, considering that rising temperatures will drive up the carbon emissions trading price, with a positive correlation between the increase in the price of primary energy and the

[1] https://www.bis.org/index.htm

increase in the carbon emissions trading price, creating the conditions for enhancing the efficiency of the current emission trading scheme (Guo et al. 2023).

To wrap up, this level of analysis is essential to highlight the important consequences of such events related to financial risks, such as credit, market, operational, liquidity and insurance risks, that create financial and price instability, where almost all central banks started to join the movement of fighting climate change to maintain financial stability (Açıkgöz & Gökmen 2022), since evidently, climate change poses new systemic risks that must be addressed within their financial stability mandate, even if the identification and management of climate-related physical and transition risks represent a reality where central banks cannot rely on traditional risk management approaches and need to explore new approaches (Bolton et al. 2020).

Second Level: Disruptive Economic Effects of CC

The transition to a low-carbon economy poses a new systemic financial risk, in which central banks need to revisit their backward-looking risk models by integrating forward-looking climate-related risks; also, future climate scenario analysis should rely on systems-based approaches with a socio-technical perspective, and embracing climate-related uncertainty by helping to create and coordinate policies to fight climate change (Svartzman et al. 2021).

Another evident disruption to the economic system is the increasing trend for strengthening disaster management capabilities, incentivizing green projects, and promotion of green finance (Ghosh et al. 2021); this is also relevant in cases where there is a need for special regulatory environments that address the concerns of a variety of stakeholders (Evans 2014).

This second level must also refer to the reality where so-called "Green Swan ideas" can counter the adverse impacts of pollution, creating opportunities for the design of a sustainable economy, based on comprehensive benefits in social, environmental and economic wealth (Posthuma et al. 2024).

Third Level: Overall Risks and the Potential Mitigation

Among the actions implemented to combat climate change are economic policy instruments, circularity models, that stem from the physical sciences, in studies of the relationship between economic activities and the natural environment from Georgescu-Roegen, based on biophysics and in particular the second law of thermodynamics and the law of entropy, constraints are imposed on economic activities (Carré 2017).

In that sense, according to Cerdá et al. (2016), the origin of circular economy lies in the field of study of industrial ecology in the early 1990s, including thoughts of functional service economics or performance economics; this concept gained

Table 9.1 Physical and transitory risk

Physical risks	Transition risks
Arise from the interaction of CC-related risks and the vulnerability of human systems and are represented as the costs and financial losses caused by the increase in the frequency and severity of climate events, which can occur in the form of storms, heat waves, floods, among other phenomena, and in the long term in climate patterns such as ocean acidification, sea level rise, droughts, among others	Represent those unexpected financial impacts that result from a change in the use of coal to clean energies caused by policies, technological advances, or provisions of social norms

Source Own elaboration based on Bolton (2020)

momentum when the Ellen MacArthur Foundation, USA, published in 2012,[2] documents supporting governments and the business sector to promote CE as a way to integrate environmental and social sustainability into economic development; according to such publications, CE is an economy that is "purposefully restorative and regenerative", a model that is not intended to have any link to the global economic development of the consumption of finite resources, where three important contributions are recognized at the level of companies and economies in terms of growth generation: job creation, reduction of the ecological footprint in the economy, and reduction of the cost of living and management in public institutions (Ellen MacArthur Foundation 2015).

According to Bolton (2020) "green swans" are the "black swans" of climate, in which the physical and transitional risks of climate change will undoubtedly materialize in the future, considering that climate catastrophes are more severe than most systemic financial crises, and that climate change is highly complex with cascading effects under an unpredictable environmental, geopolitical and economic dynamics; according to the ideas proposed by the author, there are two types of risk: physical and transitory; which are summarized in Table 9.1:

In essence, regenerative economies build lasting human, social and economic vitality through developing the richness, variety, responsibility and integrity of inclusive human networks at all levels of global civilization. The task for business, finance and policy makers is to evolve to align with regenerative principles, which give rise to healthy human networks. In other words, this system seeks to emulate or mimic natural systems that are in their entirety regenerative (Reed 2009) (Fig. 9.2).

As can be seen, the regenerative economy, which derives from regenerative capitalism as basic principles for its implementation, contemplates not only sustainability but is located in the regenerative plane looking for a natural design where a holistic and non-exclusive thinking is contemplated, motivating and encouraging the use of less energy and materials and moving away from the conventional economy.

The triple-dimension analysis helps us to understand how relevant is for practitioners, regulators and scholars to consider climate-related risks as a key priority

[2] https://www.ellenmacarthurfoundation.org/publications

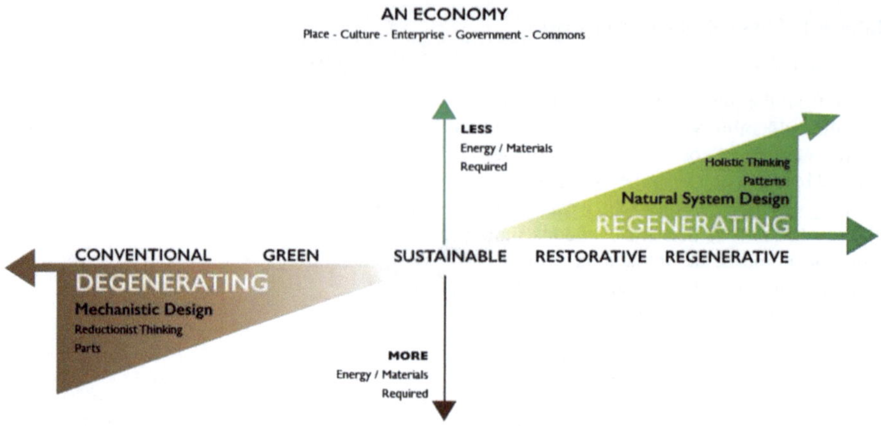

Fig. 9.2 Regenerative system. *Source* Image taken from trajectory of Ecological Design (Reed 2009)

in financial decision-making and for regulatory supervision in the financial sector, affecting a diverse group of fields such as climate science, economics, finance and risk management (Bingler and Colesanti 2022), to create a true green economy that serves the expectations of important stakeholders.

Conclusions

The green swan contemplates those aspects that derived from human activity have caused climate change and that thanks to these effects and phenomena, affect the economic and productive system of the world. On the other hand, the circular economy seeks to modify the productive systems making them sustainable and seeks as a natural consequence to avoid or mitigate those green swans that can manifest themselves suddenly affecting society and the economy in general of the countries.

So, it can be stated that the circular economy from its implementation of circular models can execute a significant change so that the green swans do not impact or do so to a lesser extent. However, in the end it is necessary that the capitalist system is modified and adapted to these new models.

The CE and green swan approaches are not exempt from criticism and risks, given the multiple interpretations and variants that can occur in the capitalist economic system, where the multiple interests of different actors keep positioned the discourses of eco-efficiency, industrial competitiveness instead of a comprehensive and global sustainability. Despite the epistemological difference between the two concepts, since the green swan theory is described as a set of factors caused by economic activity because of climate change that suddenly affects the economy as a whole, the CE corresponds rather to an economic vision that seeks through its models the achievement of sustainability.

For this reason, the inclusion of the concept of regenerative capitalism where both concepts are considered and seeks that the current capitalist system evolves if it contemplates the holistic vision where both factors are considered as important in the design and construction of an economy that lasts over time and is considered healthy.

Bibliographys

Açıkgöz Ş, Gökmen Ş (2022) Understanding the effects of green swan events on financial stability: annex II countries and Turkey. In: Modeling and advanced techniques in modern economics, pp 127–160

Banco Mundial cc (02 de septiembre de 2013). Banco Mundial. Obtenido de www.worldbank.org/en/topic/climatechange

Bingler JA, Colesanti Senni C (2022) Taming the Green Swan: a criteria-based analysis to improve the understanding of climate-related financial risk assessment tools. Climate Policy 22(3):356–370. https://doi.org/10.1080/14693062.2022.2032569

Bolton P, Després M, da Silva LAP, Samama F, Svartzman R (2020) What role for central banks in the age of climate-related risks and other "Green Swans"? Regards Croises Sur L'economie 26(1):110–122

Bolton P (2020) The green swan. Central Banking and financial stability in the age of climate change . Banco internacional de pagos

Carré JG (2017) Does air pollution play a role in infertility? a systematic review. Environ Health 16:1–16

Cerdá E (2016) Economía circular. Economía Industrial 401(3):11–20

Chaves Ávila R (2018) La economía social ante los paradigmas económicos emergentes: innovación social, economía colaborativa, economía circular, responsabilidad social empresarial, economía del bien común, empresa social y econo. CIRIEC-España Revista de economía pública, social y cooperativa, 5–50

CMNUCC (1997) Convención Marco de las Naciones Unidas sobre el Cambio Climático. Naciones Unidas, Montreal

Coeure B (2020) Monetary Policy and Climate Change. European Central Bankl, pp 210–222

Conde C (2007) México y el cambio climático global. Dirección general de divulgación de ciencias, México

Cordero G (2012) El cambio Climático. Ciencia y Sociedad 227–240

Ellen MacArthur Foundation (2015) Delivering the circular economy. Obtenido de toolkit for policymakers. https://ellenmacarthurfoundation.org/a-toolkit-for-policymakers

Elliott C (2023) 8 Whitewashing green swans. Uncertainty in Global Politics, 136

Estrada Porrúa M (2001) Cambio climático global: causas y consecuencias. Datos Hechos y lugares, 8–11

Evans L (2014) Recent judgments on the health claims regulation: a journey through the colourful world of health claims made on food stopping at Luxembourg and Karlsruhe. Eur Food Feed Law Rev 233–240

Ghosh S, Kundu S, Dilip A (2021) Green swans and their economic impact on Indian Coastal States. Reserve Bank of India Occasional Papers 42(1):1–69

Guo Y, Li Y, Wu H, Hao Y (2023) Carbon emission trading under the wings of black swans and green swans: Evidence from China. Int J Finance Economics

Iberdrolla (2020) Infografía sobre Efectos Socioeconomicos del cambio climático. Obtenido de https://www.Iberdrola.com: https://www.Iberdrola.com/wcorp/gc/prod/es_ES/comunicación/docs/Infografia
IPCC (2007) CAMBIO CLIMÁTICO 2007 INFORME DE SÍNTESIS. OMM, Ginebra, Suiza
Lopez Ev (2007) cambio climatico y cohesión social. En E. V. Lopez. ur-ball III, Barcelona
Posthuma L, Bloor M, Campos B, Groh K, Leopold A, Sanderson H, Thomas P (2024) Green Swans countering chemical pollution. Integr Environ Assess Manag 20(3):888–891
Reed B (2009) The integrative design guide to green building: Redefining the practice of sustainability. John Wiley & Sons (Vol. 43)
Sahakian M (2016) The Social and Solidarity Economy: Why is it relevant to Industrial Ecology? En Clift R, Druckman A. Taking. Stock of Industrial Ecology, 205–227
Samuelson PA (2006) Economía. McGraw-Hill 25a edición, España
Singhania M, Gupta S, Chadha G, Braune E, Dana LP, Idowu SO (2023) Mapping 26 years of climate change research in finance and accounting: a systematic scientometric analysis. Environ Sci Pollut Res 30(35):83153–83179
Stern N (2016) The economics of climate change. Cambridge University Press
Svartzman R, Bolton P, Despres M, Pereira Da Silva LA, Samama F (2021) Central banks, financial stability and policy coordination in the age of climate uncertainty: a three-layered analytical and operational framework. Climate Policy 21(4):563–580
Taleb NN (2007) Black swans and the domains of statistics. Am Stat 61(3):198–200
Terceiro Lomba J (2018) Economia del Cambio Climático. Taurus, España

Chapter 10
Competitiveness of Decent Work of the US Textile Sector Using a Multivariable Classification of the SDG 8 Target 5: Full Employment and Decent Work

María de la Soledad del Río, Nadyra Rodríguez Arias, César Gustavo Iriarte Rivas, and Elitania Leyva Rayón

Introduction

The 2030 Agenda for Sustainable Development serves as a collective roadmap for promoting prosperity and sustainability with a global perspective, which stand as an urgent call to action for every nation, institution, organization and individual regardless of its development status; it recognizes the imperative of addressing poverty and various forms of deprivation in conjunction with efforts to enhance healthcare and education, diminish inequality, and stimulate sustainable economic growth as well as decent work, encompassing critical domains such as water, energy, climate, oceans, urbanization, transport, science and technology.[1]

In such scenario, the textile industry serves as one of the manufacturing sectors that emerge in economically disadvantaged nations, where it is frequent to see employment opportunities to individuals with limited skills, who have limited options for paid work; for those employed, particularly women, generally textile manufacturing serves as a pathway to gradually escape poverty, where the predominance of women represents a crucial stride toward poverty alleviation for such social sector.[2]

The targets for the Social Development Goal 8 "Promote inclusive and sustainable economic growth, employment and decent work for all", are designed to ensure decent work and sustainable economic growth, including factors such as

[1] https://sdgs.un.org/goals.

[2] https://textilesforsdgs.org/sdgs/goals/8-decent-work-and-economic-growth/.

M. de la Soledad del Río (✉) · N. R. Arias · C. G. I. Rivas · E. L. Rayón
Arkansas State University Campus Querétaro, Querétar, QUE, Mexico
e-mail: mdelrio@astate.edu

A. E. Toscano-Hernández et al. (eds.), *Sustainability, Stakeholders and Marketing in the Textile Sector*, SDGs and Textiles, https://doi.org/10.1007/978-981-97-6952-0_10

sustainable economic growth (Sustain per capita economic growth in accordance with national circumstances), diversification, innovation and productivity (higher levels of economic productivity through diversification, technological upgrading and innovation), promotion of policies to support job creation and growing enterprises (supporting productive activities, decent job creation, entrepreneurship, creativity and innovation, formalization and growth of micro-, small- and medium-sized enterprises), as well as improving resource efficiency in consumption and production.

In such scenario, the SDG 8 underscores the need for economic expansion to serve as a beneficial global force, by promoting an economic progress that generates meaningful and sustainable employment opportunities without causing harm to the environment, safeguarding labor rights and eradicating modern slavery and child labor are essential objectives and by encouraging job generation and enhancing access to financial services.[3]

Literature Review

According to some authors, there are six SDG approaches that represent the essential building-blocks of achievement in the practical implementation of the development goals, that ranges from education, gender and inequality to health, well-being and demography; energy decarbonization and sustainable industry; sustainable food, land, water and oceans; sustainable cities and communities; and digital revolution for sustainable development, in where every aspect identifies calling for actions by well-defined stakeholders (Sachs et al. 2019).

Thus, putting SDGs ideas to work trying solving a variety of social issues from poverty and hunger alleviation to sustainable cities, economies, and ecosystems represents a multi-sectoral, holistic and complex approach, since governments, institutions, organizations, society and other relevant stakeholders are looking forward to secure the well-being and sustainability to both current and future generations, and thus, national strategies must be built on sound science and engagement of local stakeholders (Wood et al. 2018).

To achieve such ambitious plan, the new Sustainable Development Goals and their targets are designed to guide the decisions throughout the next fifteen years, by exerting an influence in the international politics and the finance available for sustainable development, to fight extreme poverty, seizing the potential for the SDGs to mobilize academic communities, professional practice communities and social movements to demand relational change and liability for the decision makers (Gusmão Caiado et al. 2018), where the SDGs represents a practical guide to end poverty, protect the planet, and ensure prosperity for all (Leal Filho et al. 2018).

The SDG 8, which is related to the promotion of sustained, inclusive and sustainable economic growth by providing full and productive employment and decent work

[3] https://www.globalgoals.org/goals/8-decent-work-and-economic-growth/#things-to-do.

shows that responsible and sustainable practices are important not only for the particular goals, but also for many of the other targets under different SDGs, (SDG 8 aiming for economic growth and SDG 9 promoting industrialization and infrastructure are logical examples of such relation) (Schroeder et al. 2019).

Decent Work

Considering decent work as a condition where a given worker is performing a remunerate productive activity under adequate working conditions (Mariano et al. 2023), in a context that represents sound job opportunities and reduced level of unemployment (Finatto et al. 2023), the reality shows that in a number of economies one of the most pressing challenges relate to working poverty, informality, and underemployment as well (Barford et al. 2021).

In that sense, the Sustainable Development Goal (SDG) 8 "Decent Work and Economic Growth" was particularly vulnerable to the multidimensional crisis caused by COVID19, including challenges such as lockdowns, unemployment, and financial flows reorientation, among others (Makarenko et al. 2021); considering that such goal promotes a sustained, inclusive and sustainable economic growth, with full and productive employment and decent work for all, some researchers are highlighting the relevance of understanding productive employment and decent work considering the value and costs of social reproduction, in where both gender and labour rights must underpin SDG 8 to promote a real inclusive, sustainable and decent work (Rai et al. 2019).

In what it comes to the key components of SDG 8, cost of living, equity, safety, security, integrity and well-being in the workplace, and the corresponding relationship with economic growth represents a real challenge for implementation in the reality of industries and organizations, finding some issues that arise in the reality of workplace, such as weak treatment of employees, safe and secured working environment and low promotion of wellbeing and integrity at the workplace (Ruxho et al. 2023).

To clearly identify the constructs related to achieving a sustained and inclusive economic growth, any socioeconomic system must address structural injustices underlying inequalities, and exploitative labor practices, by fostering the notion of sustained and inclusive growth that put the human wellbeing at the center of the economic design (Bianchi and de Man 2021); while the 2030 Agenda embraces market-centred institutional arrangements grafting the human rights to full employment and decent work onto a business-oriented economic growth agenda, the ambiguity in SDG 8 remains a challenge to be solved (Frey 2018).

Thus, some study cases offer insight related to decent work, job engagement and turnover intention in the retail, where rapid changes in the industry have led to a deterioration and misunderstanding of working standards, since maximizing labor force potential, job engagement and turnover intentions are considered mainly in economic terms (Chada et al. 2022); some other researchers highlight lacking of flexible work systems, scarcity of gender-sensitive organizational policies and labor policies, remain as some of the main difficulties faced in terms of implementing decent work policies (Kemechian et al. 2023).

The Theoretical Approach to Decent Work Competitiveness

In order to be able to build a proper construct related to competitiveness, it is important to address some questions regarding the understanding of such concept in the economic theory; in that sense, it is essential to recognize the diversity of contexts and level of competitiveness as well, fundamentally measured using a quantitative approach by identifying financial success as an indicator of comparison relative of the industry rivals (Porter 1991).

Thus, following this understanding of the nature of this concept, competitiveness represents a contrast comparison measurement in nature, by identifying leaders in terms of financial success, at a diversity of levels: individual, organizational, regional, and national; in such comparison analysis, how competitive a given subject is, represents a categorical variable which output is quantitative, and it is strongly linked to financial output; but given the recent events in the global competitive environment, there is an evident need to consider how this understanding is adequate to address social issues, as in regions that have people with fading skills and poor education (Esparza 2018).

Following that line of though, studying the potential influence of fair employment practices and financial inclusion on economic growth reveal that conducive environments for decent work and enhanced access to digital financial services positively impact economic growth, which reveals a synergistic effect of a diversity of the SDG 8 fundamental components in high-income countries (Zehri et al. 2024), making evident the need to understand competitiveness in decent work as a comprehensive indicator related to the overall environment of a given industry and region.

Next, in what it comes to the idea of designing a comprehensive indicator related to decent work competitiveness, this approach must address the environmental quality including issues related to environmental degradation, sustained economic growth, clean and affordable energy, and quality education (Sinha et al. 2020).

Considering the previous context, the present work aims to provide more insights regarding decent work in the textile industry as a multidimensional construct that represents a broader view of the conditions needed to understand decent work according to the ideals of the United Nations SDG´s, and the reality of the textile industry in United States of America; with that purpose, the research question that guides the research is at follows:

- *How significant are the GDP per American state, level of education, safe working environment. use of clean energy and unemployment rate, to explain decent work competitiveness as a complex construct in the textile industry of USA?*

Method and Data

The present research uses a machine learning model by supervised learning to analyze quantitative data using as a primary source the U.S. Bureau of Labor Statistics,[4] which consist of an artificial network to analyze the data related to a variable model as it is described in Table 10.1; the dependent variable represents a multi-categorical variable consisting of 3 different outputs related to decent work competitiveness in the textile industry of United States.

For that purpose, we are considering decent work competitiveness as an ordinal variable based on a comparison to industry standards, with at least 3 competitive states: Competitive advantage (superior to the industry/region/national average), Competitive (equal to the industry/region/national average), non-competitive (less than the industry/region/national average), as follows:

Table 10.1 shows a variable model with the dependent and independent variables, including 3 possible states for the ordinal variable (based on the comparison of wages at industry and regional levels), considering the data available in the Occupational Employment and Wage Statistics,[5] which can be represented graphically as follows:

Figure 10.1 shows a graphical network of the variable model, including some descriptive relations among the considered variables of the problem, with the independent variables (including labour rights, employment, energy, economy and

Table 10.1 Variable model

Dependent variable	Type of variable	Measurement unit	Source
Decent work competitiveness	Categorical (Competitive advantage)	Wages compared to the industry/region average (>Industry average = Competitive advantage)	https://www.bls.gov/oes/tables.htm
Independent variable	**Type of variable**	**Measurement unit**	**Source**
Economy	Quantitative	GDP per state, per region	https://www.bea.gov/data/gdp/gdp-state
Education	Quantitative	College completion rates 1970–2021 for adults 25 years and older	https://data.ers.usda.gov/reports.aspx?ID=17829
Labour rights	Quantitative	Safe working environment	https://www.bls.gov/iif/state-data.htm#NTL
Energy	Quantitative	Use of clean energy	https://www.eia.gov/renewable/state/archive/2010/
Employment	Quantitative	Unemployment/employment rate	https://www.bls.gov/web/laus/laumstrk.htm

Source Own elaboration (2023)

[4] https://www.bls.gov/oes/tables.htm.

[5] https://www.bls.gov/oes/tables.htm.

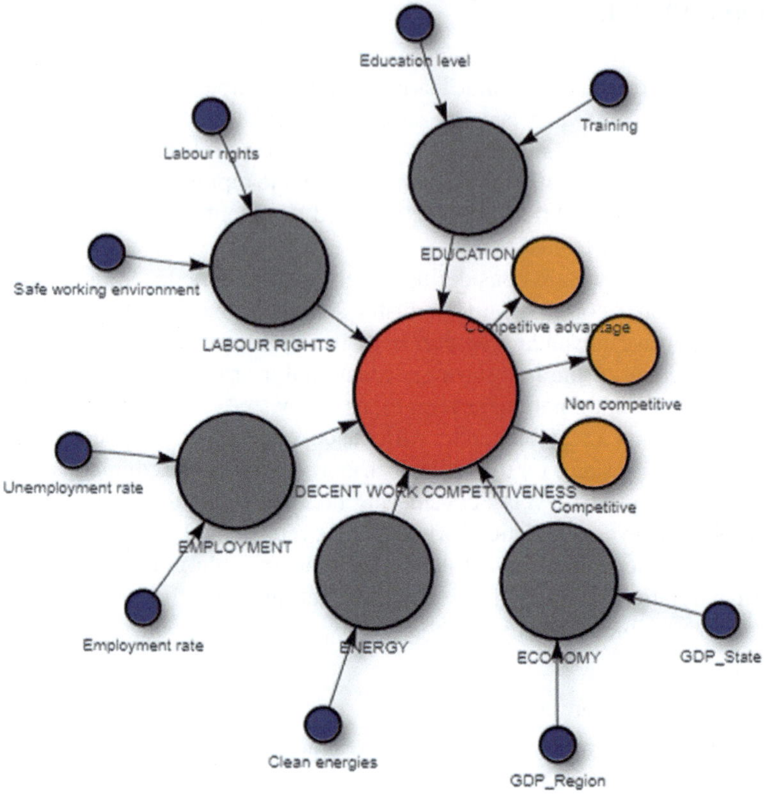

Fig. 10.1 Network analysis representation of Decent Work Competitiveness. *Source* Own elaboration with Rstudio (2023)

education) having an influence on the ordinal variable decent work competitiveness, as a description of the proposed network relationship to understand decent work accordingly to the SDG-8.

Results

Productive Activities in the Industries Related to Textiles in USA

In order to perform the present analysis, we included the following productive activities as part of the employment in the textile sector: Pressers, Textile, Garment, and Related Materials; Textile Bleaching and Dyeing Machine Operators and Tenders; Textile Cutting Machine Setters, Operators, and Tenders; Textile Knitting and

Weaving Machine Setters, Operators, and Tenders; Textile Winding, Twisting, and Drawing Out Machine Setters, Operators, and Tenders; and Textile, Apparel, and Furnishings Workers.

GDP Per State, Per Region

See Table 10.2.

Table 10.2 GDP per state with data of 2022, including the difference

State	GDP	Difference with national average	Percentage of difference with national average (%)
Alabama	235,807.30	−189,103.98	−44.50
Alaska	50,315.10	−374,596.18	−88.16
Arizona	403,474.20	−21,437.08	−5.05
Arkansas	137,355.50	−287,555.78	−67.67
California	3,167,460.80	2,742,549.52	645.44
Colorado	416,114.40	−8,796.88	−2.07
Connecticut	276,668.80	−148,242.48	−34.89
Delaware	75,172.60	−349,738.68	−82.31
District of Columbia	144,029.60	−280,881.68	−66.10
Florida	1,218,430.20	793,518.92	186.75
Georgia	655,827.30	230,916.02	54.34
Hawaii	85,211.40	−339,699.88	−79.95
Idaho	91,683.70	−333,227.58	−78.42
Illinois	864,170.90	439,259.62	103.38
Indiana	396,009.20	−28,902.08	−6.80
Iowa	197,846.30	−227,064.98	−53.44
Kansas	174,794.60	−250,116.68	−58.86
Kentucky	217,568.10	−207,343.18	−48.80
Louisiana	231,262.30	−193,648.98	−45.57
Maine	72,413.50	−352,497.78	−82.96
Maryland	412,282.60	−12,628.68	−2.97
Massachusetts	604,357.70	179,446.42	42.23
Michigan	539,898.20	114,986.92	27.06
Minnesota	379,111.50	−45,799.78	−10.78
Mississippi	114,152.90	−310,758.38	−73.13
Missouri	336,625.80	−88,285.48	−20.78

(continued)

Table 10.2 (continued)

State	GDP	Difference with national average	Percentage of difference with national average (%)
Montana	53,983.40	−370,927.88	−87.30
Nebraska	137,078.10	−287,833.18	−67.74
Nevada	187,226.20	−237,685.08	−55.94
New Hampshire	90,150.50	−334,760.78	−78.78
New Jersey	646,731.30	221,820.02	52.20
New Mexico	101,315.00	−323,596.28	−76.16
New York	1,763,524.60	1,338,613.32	315.03
North Carolina	609,058.20	184,146.92	43.34
North Dakota	54,798.70	−370,112.58	−87.10
Ohio	689,681.30	264,770.02	62.31
Oklahoma	191,583.00	−233,328.28	−54.91
Oregon	254,708.40	−170,202.88	−40.06
Pennsylvania	772,335.90	347,424.62	81.76
Rhode Island	62,190.90	−362,720.38	−85.36
South Carolina	250,873.40	−174,037.88	−40.96
South Dakota	54,959.30	−369,951.98	−87.07
Tennessee	412,101.30	−12,809.98	−3.01
Texas	1,924,007.50	1,499,096.22	352.80
Utah	213,898.30	−211,012.98	−49.66
Vermont	34,609.40	−390,301.88	−91.85
Virginia	576,964.30	152,053.02	35.78
Washington	641,143.60	216,232.32	50.89
West Virginia	76,525.50	−348,385.78	−81.99
Wisconsin	335,688.50	−89,222.78	−21.00
Wyoming	37,294.20	−387,617.08	−91.22

Source Own elaboration with data from Bureau of Economic Analysis (n.d.)[6]

Mean Annual Wage

For what it comes to the mean annual wage, (which is the estimated mean hourly wage of an occupation multiplied by 2,080 h), the average in terms of wages is 29,807 USD, for all the USA states; the highest wages in such industry are registered in Maryland ($39,550), Massachusetts ($38,670), California ($35,640), Maine ($35,410), and Washington ($35,020). Conversely, the lowest wages registered in such industry are Wyoming ($25,420), Arkansas ($25,230), Texas ($25,060), Alabama ($24,900), and Mississippi ($23,700), as it can be seen in the Table 10.3.

[6] https://www.bea.gov/data/gdp/gdp-state.

Table 10.3 Mean annual wage, (which is the estimated mean hourly wage of an occupation multiplied by 2,080 h) per USA state in the pressers, textile, garment, and related materials industries (Year 2022)

State	Mean annual wage
Maryland	39,550
Massachusetts	38,670
California	35,640
Maine	35,410
Washington	35,020
New York	34,540
Connecticut	33,810
Rhode Island	33,570
Louisiana	33,350
Illinois	33,020
Wisconsin	32,870
Oregon	32,860
Idaho	31,880
New Jersey	31,600
Colorado	31,030
Minnesota	30,940
New Hampshire	30,340
Michigan	30,020
Arizona	29,850
Nebraska	29,050
Missouri	28,750
Virginia	28,690
Florida	28,640
Utah	28,490
Kentucky	28,200
New Mexico	28,190
Ohio	28,020
Pennsylvania	28,020
Nevada	27,900
Indiana	27,430
Tennessee	27,160
Iowa	27,140
Georgia	26,680
North Carolina	26,400

(continued)

Table 10.3 (continued)

State	Mean annual wage
Kansas	26,360
South Dakota	26,350
Oklahoma	26,040
South Carolina	25,890
Wyoming	25,420
Arkansas	25,230
Texas	25,060
Alabama	24,900
Mississippi	23,700

Source Own elaboration based on the Occupational Employment and Wage Statistics (n.d.)[7]

According to the previous information, the annual average is $28,807; using a comparative descriptive analysis, the competitiveness of this variable is referred in the Table 10.4 (using a $\pm 10\%$ radio in the difference comparison per state, divided by the average mean).

Table 10.4 Level of decent work competitiveness per American State

State	Comparative percentage in terms of the difference with the average (%)	Competitiveness
Maryland	25	Competitive advantage
Massachusetts	23	Competitive advantage
California	16	Competitive advantage
Maine	16	Competitive advantage
Washington	15	Competitive advantage
New York	14	Competitive advantage
Connecticut	12	Competitive advantage
Rhode Island	11	Competitive advantage
Louisiana	11	Competitive advantage
Illinois	10	Competitive advantage
Wisconsin	9	Competitive
Oregon	9	Competitive
Idaho	7	Competitive
New Jersey	6	Competitive
Colorado	4	Competitive
Minnesota	4	Competitive

(continued)

[7] https://www.bls.gov/oes/current/oes_nat.htm.

Table 10.4 (continued)

State	Comparative percentage in terms of the difference with the average (%)	Competitiveness
New Hampshire	2	Competitive
Michigan	1	Competitive
Arizona	0	Competitive
Nebraska	−3	Competitive
Missouri	−4	Competitive
Virginia	−4	Competitive
Florida	−4	Competitive
Utah	−5	Competitive
Kentucky	−6	Competitive
New Mexico	−6	Competitive
Ohio	−6	Competitive
Pennsylvania	−6	Competitive
Nevada	−7	Competitive
Indiana	−9	Competitive
Tennessee	−10	Competitive
Iowa	−10	Competitive
Georgia	−12	Not competitive
North Carolina	−13	Not competitive
Kansas	−13	Not competitive
South Dakota	−13	Not competitive
Oklahoma	−14	Not competitive
South Carolina	−15	Not competitive
Wyoming	−17	Not competitive
Arkansas	−18	Not competitive
Texas	−19	Not competitive
Alabama	−20	Not competitive
Mississippi	−26	Not competitive

Source Authors (2024)

As it can be seen in the Table 10.4, the data shows that 10 states has competitive advantage, showing a difference expressed in percentages more than 10 percent in comparison to the average; 22 are competitive, with a ±10% comparative difference with the average on the Country, and the remaining 11 are not competitive in that same logic, as it can be seen in Fig. 10.2.

The figure above shows a comparison among the American states in terms of differences among percentages in relation to the national average, and the situation of every state in relation to a level of competitiveness; in that sense, the percentages

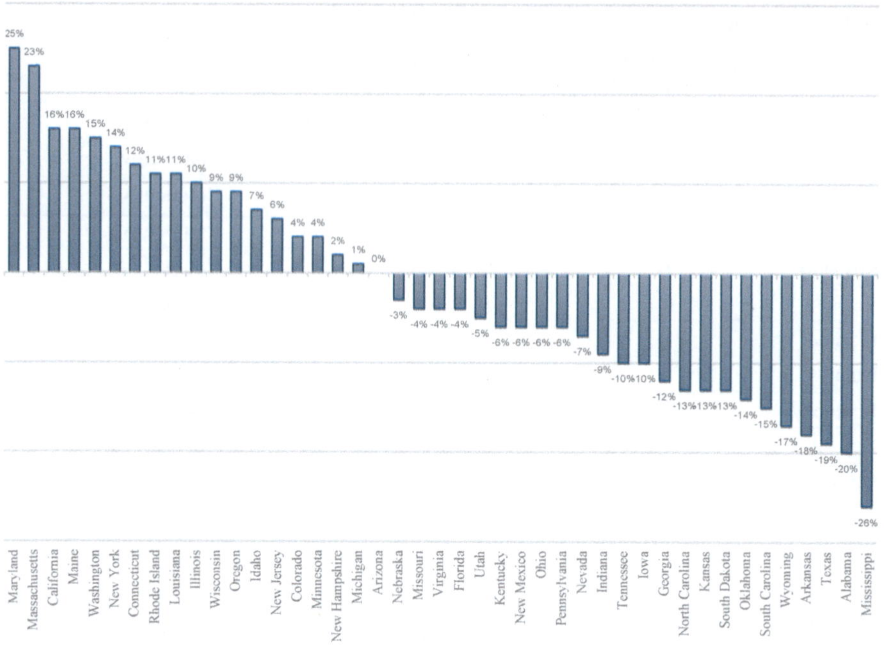

Fig. 10.2 Comparative analysis in relation to the level of competitiveness in Wages per USA state. *Source* Authors (2024)

depicted in the graphic represents the descriptive analysis of the dependent categorical variable "Competitiveness" using an ordinal logic (Competitive, Not Competitive and Competitive Advantage), thus, by identifying every state, the following analysis will include the corresponding quantitative analysis for the present data.

Level of Education

The next figure represents the percentage of total population that completed college (including urban and rural) in the United States, per state, showing the completion rates in the period 2017–2021 for adults 25 years and older, as can be seen in the Fig. 10.3.

The previous figure can also be represented in a Table 10.5 that tracks an evolution towards a period of time considering 1970 to 2021, generally showing increasing numbers per American State, as well as the final percentage including a period of time in the 4 last years, as the following Table 10.5.

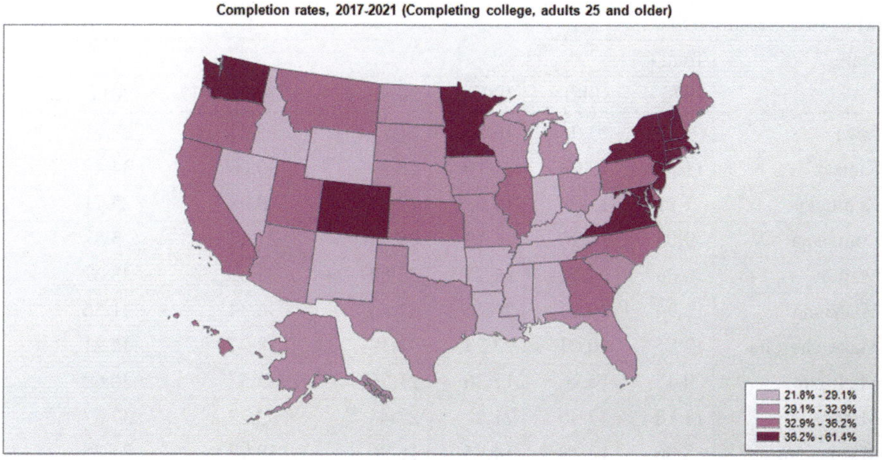

Fig. 10.3 Total population that completed college (including urban and rural) in the United States, per State. *Source* Own elaboration with data from USDA (n.d.)[8]

Table 10.5 Completion rates 1970–2021 for adults 25 years and older

State	Total (%)					
	1970	1980	1990	2000	2008–2012	2017–2021
United States	10.66	16.23	20.34	24.40	28.48	33.67
Alabama	7.85	12.18	15.68	19.03	22.26	26.67
Alaska	14.14	21.08	23.03	24.71	27.48	30.59
Arizona	12.56	17.41	20.29	23.53	26.63	31.21
Arkansas	6.66	10.82	13.32	16.66	19.80	24.30
California	13.44	19.60	23.36	26.62	30.45	35.27
Colorado	14.93	22.96	26.97	32.69	36.68	42.83
Connecticut	13.66	20.66	27.18	31.41	36.16	40.55
Delaware	13.09	17.47	21.41	25.05	28.52	33.65
District of Columbia	17.78	27.46	33.31	39.07	51.21	61.40
Florida	10.26	14.88	18.28	22.33	26.17	31.53
Georgia	9.21	14.59	19.32	24.30	27.76	33.00
Hawaii	13.97	20.35	22.88	26.17	29.61	34.34
Idaho	9.96	15.83	17.65	21.67	24.71	29.13
Illinois	10.31	16.20	21.04	26.06	31.08	36.16
Indiana	8.33	12.52	15.56	19.41	22.98	27.84

(continued)

[8] https://data.ers.usda.gov/reports.aspx?ID=17829.

Table 10.5 (continued)

State	Total (%)					
	1970	1980	1990	2000	2008–2012	2017–2021
Iowa	9.10	13.94	16.85	21.21	25.32	29.71
Kansas	11.38	17.03	21.10	25.80	30.00	34.40
Kentucky	7.21	11.08	13.63	17.14	20.98	25.71
Louisiana	9.05	13.90	16.13	18.73	21.42	25.53
Maine	8.36	14.40	18.77	22.87	27.32	33.62
Maryland	13.94	20.36	26.46	31.45	36.34	41.56
Massachusetts	12.55	20.03	27.23	33.19	38.97	45.21
Michigan	9.42	14.26	17.36	21.76	25.53	30.62
Minnesota	11.10	17.36	21.82	27.43	32.23	37.57
Mississippi	8.06	12.29	14.75	16.90	19.97	23.25
Missouri	8.97	13.89	17.80	21.58	25.77	30.69
Montana	11.01	17.47	19.79	24.37	28.48	33.66
Nebraska	9.65	15.52	18.94	23.74	28.11	32.95
Nevada	10.79	14.44	15.28	18.16	22.22	26.06
New Hampshire	10.85	18.21	24.37	28.65	33.44	38.23
New Jersey	11.84	18.34	24.85	29.78	35.38	41.50
New Mexico	12.73	17.58	20.41	23.45	25.63	28.53
New York	11.89	17.94	23.13	27.37	32.79	38.11
North Carolina	8.45	13.24	17.38	22.46	26.83	33.02
North Dakota	8.39	14.82	18.07	21.99	27.06	31.09
Ohio	9.27	13.65	16.97	21.10	24.73	29.69
Oklahoma	10.02	15.07	17.79	20.28	23.23	26.85
Oregon	11.78	17.85	20.60	25.08	29.25	35.04
Pennsylvania	8.75	13.56	17.94	22.35	27.05	33.11
Rhode Island	9.39	15.43	21.27	25.60	30.76	35.32
South Carolina	8.97	13.42	16.65	20.42	24.56	29.84
South Dakota	8.60	14.01	17.16	21.51	26.05	29.99
Tennessee	7.89	12.60	15.96	19.56	23.49	29.04
Texas	10.87	16.88	20.32	23.24	26.26	31.53
Utah	13.98	19.88	22.26	26.13	29.86	35.43
Vermont	11.49	19.04	24.31	29.45	34.21	40.86
Virginia	12.30	19.09	24.48	29.46	34.68	40.33
Washington	12.72	18.95	22.93	27.73	31.57	37.28
West Virginia	6.77	10.42	12.33	14.83	17.91	21.78

(continued)

Table 10.5 (continued)

State	Total (%)					
	1970	1980	1990	2000	2008–2012	2017–2021
Wisconsin	9.76	14.83	17.74	22.42	26.40	31.49
Wyoming	11.78	17.15	18.79	21.91	24.26	28.52
Puerto Rico				0.182777	0.224841	0.274314

Source Own elaboration with data from USDA (2024)[9]

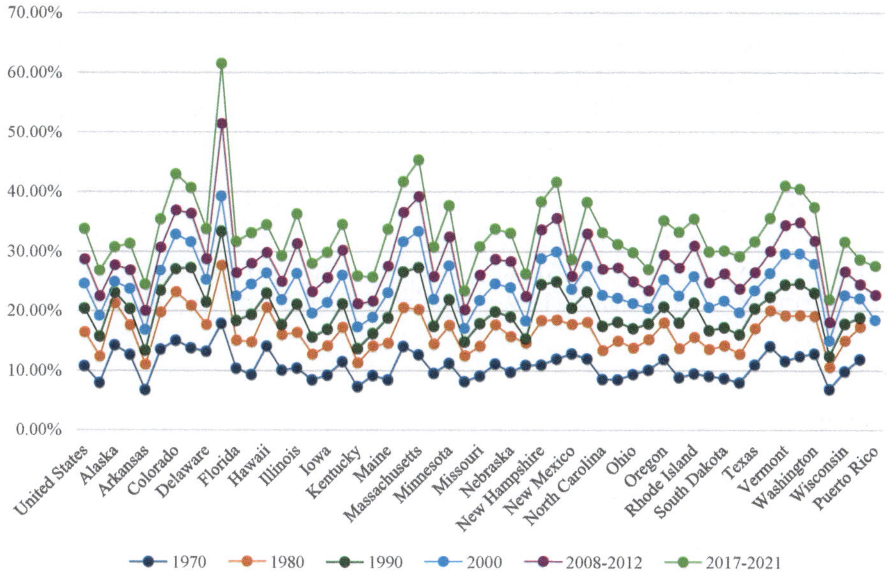

Fig. 10.4 Contrasting figures including a time series analysis per State and Year for completion rates. *Source* Own elaboration with information from USDA (n.d.)[10]

The trends present on the Table 10.5 are depicted in the following figure, as follows.

The Fig. 10.4 represents the contrast per American state, considering a time lapse from 1970 to 2021, showing a consistent trend per individual state, with positive increasing numbers in every period considered, having the 2017–2021 lapse as the highest.

[9] https://data.ers.usda.gov/reports.aspx?ID=17829.

[10] https://data.ers.usda.gov/reports.aspx?ID=17829.

Safe Working Environment

Safe working environment regards healthy and safe working conditions where the workers are performing their activities in an environment that protects them from potential workplace injuries, illnesses, and fatalities; in that sense the Survey of Occupational Injuries and Illnesses (SOII) contributes with useful information since meticulously gather and analyze data pertaining to variables related to safe working conditions, offering data that shows crucial details such as the nature and severity of the injury or illness, the body parts affected, the duration of time lost from work, and the specific industry sector where the incident occurred; for what it comes to the USA states, the total recordable cases of nonfatal occupational injuries and illnesses in the private industry are showed below.

As it is shown in the Table 10.6, the states that record the highest number of nonfatal occupational injuries are California (363.8), Texas (178), New York (140.2), Pennsylvania (135.4), Illinois (110), Ohio (93.1), Michigan (90.9), Washington (89.3), Georgia (77.5), and New Jersey (74.4); conversely, the lowest record for nonfatal occupational injuries and illnesses by state are Virgin Islands (0.3), Guam (1.3), Wyoming (5.4), Alaska (7.1), Delaware (7.5), District of Columbia (7.8), Vermont (9.6), Montana (11.9), Hawaii (13.4), and West Virginia (13.6).

Use of Clean Energy

The dynamic nature of the U.S. energy landscape is engaged in an ongoing evolution towards cleaner and more sustainable energy sources; in the past, the landscape of electricity generation in the United States was predominantly characterized by the utilization of fossil fuels, encompassing coal, natural gas, and petroleum; in the recent years, there has been a gradual shift towards greater reliance on natural gas and renewables, while the prominence of coal-fired electricity has steadily diminished; this reality has been powered by the ascendance of natural gas with a share of electricity-generation capacity soaring from 17% in 1990 to a 43% in 2022. Likewise, nuclear and hydropower facilities in the United States have maintained a relatively stable position, consistently representing around 20% of the nation's annual electricity generation.[11]

In that sense, the Table 10.7 shows the descriptive comparison among USA states in terms of net generation, for gigawatts and gigawatts-hours in terms of sustainable energy sources, aggregating solar, geothermal, hydroelectric, wind and biomass.

As Table 10.7 shows, the American States leaders in clean energy generation are Washington (23.884), California (16.46), Oregon (10.684), Texas (10.985), New York (6.033), Alabama (3.855), Iowa (3.728), Montana (3.085), Idaho (3.14), and Arizona (2.901); on the other hand, the States with the lower generation of green

[11] https://www.eia.gov/energyexplained/electricity/electricity-in-the-us-generation-capacity-and-sales.php.

Table 10.6 Numbers of nonfatal occupational injuries and illnesses by state and case type, private industry (thousands)

State	Total recordable cases
Private industry[a]	2834.5
Alabama	36.5
Alaska	7.1
Arizona	58.4
Arkansas	19.6
California	363.8
Connecticut	37.2
Delaware	7.5
District of Columbia	7.8
Georgia	77.5
Guam	1.3
Hawaii	13.4
Illinois	110
Indiana	69.9
Iowa	35.9
Kansas	29.2
Kentucky	41.2
Louisiana	25.6
Maine	19.1
Maryland	50
Massachusetts	65.7
Michigan	90.9
Minnesota	62.9
Missouri	53.6
Montana	11.9
Nebraska	21.7
Nevada	34.7
New Jersey	74.4
New Mexico	14.9
New York	140.2
North Carolina	73.4
Ohio	93.1
Oregon	47.4

(continued)

Table 10.6 (continued)

State	Total recordable cases
Pennsylvania	135.4
Puerto Rico	17.6
South Carolina	33.7
Tennessee	60.6
Texas	178
Utah	28.2
Vermont	9.6
Virgin Islands	0.3
Virginia	65
Washington	89.3
West Virginia	13.6
Wisconsin	70.4
Wyoming	5.4

Days-away-from-work cases include those that result in days away from work with or without job transfer or restriction

[a] Data cover all 50 states

Note Because of rounding, components may not add to totals. Dash indicates data do not meet publication guidelines

Source U.S. Bureau of Labor Statistics, U.S. Department of Labor

Table 10.7 Net generation, for gigawatts and gigawatts-hours in terms of sustainable energy sources, aggregating solar, geothermal, hydroelectric, wind and biomass

State	Net summer capacity (gigawatts)	Rank	Net generation (gigawatt-hours)	Rank
Alabama	3.855	6	11.081	6
Alaska	0.422	41	1.452	42
Arizona	2.901	10	6.941	14
Arkansas	1.667	21	5.283	18
California	16.46	2	58.881	2
Colorado	2.01	17	5.133	19
Connecticut	0.281	45	1.13	44
Delaware	0.01	50	0.138	50
Florida	1.182	29	4.664	21
Georgia	2.689	12	6.502	16
Hawaii	0.34	44	0.817	48
Idaho	3.14	9	10.168	7

(continued)

Table 10.7 (continued)

State	Net summer capacity (gigawatts)	Rank	Net generation (gigawatt-hours)	Rank
Illinois	2.112	19	5.257	29
Indiana	1.452	28	3.699	38
Iowa	3.728	7	10.309	10
Kansas	1.082	30	3.473	33
Kentucky	0.893	31	3.02	28
Louisiana	0.517	38	3.577	30
Maine	1.692	22	7.963	11
Maryland	0.799	34	2.241	34
Massachusetts	0.566	39	2.27	35
Michigan	0.807	33	4.083	25
Minnesota	2.588	14	7.48	12
Mississippi	0.235	46	1.504	41
Missouri	1.03	32	2.527	36
Montana	3.085	8	10.442	9
Nebraska	0.443	43	1.807	47
Nevada	1.507	24	4.444	23
New Hampshire	0.671	35	2.71	32
New Jersey	0.23	47	0.868	46
New Mexico	0.818	36	2.072	40
New York	6.033	5	32.286	4
North Carolina	2.449	13	6.84	13
North Dakota	1.941	20	6.15	22
Ohio	0.231	48	1.129	45
Oklahoma	2.412	15	6.969	15
Oregon	10.684	3	35.299	3
Pennsylvania	1.984	16	6.577	17
Rhode Island	0.028	49	0.144	49
South Carolina	1.623	23	4.25	24
South Dakota	2.223	18	6.611	20
Tennessee	2.847	11	9.125	8
Texas	10.985	4	28.967	5
Utah	0.528	40	1.476	43
Vermont	0.408	42	1.829	39

(continued)

Table 10.7 (continued)

State	Net summer capacity (gigawatts)	Rank	Net generation (gigawatt-hours)	Rank
Virginia	1.487	26	3.72	26
Washington	23.884	1	74.905	1
West Virginia	0.715	37	2.307	37
Wisconsin	1.267	27	4.586	27
Wyoming	1.722	25	4.271	31

Source Own elaboration with data from U.S. Energy Information Administration (n.d.)[12]

energy are Delaware (0.01), Rhode Island (0.028), Ohio (0.231), New Jersey (0.23), Mississippi (0.235), Connecticut (0.281), Hawaii (0.34), Nebraska (0.443), Vermont (0.408), Alaska (0.422), and Utah (0.528).

Unemployment/Employment Rate

To gather the information regarding unemployment rate, we visited the official website from the U.S. Bureau Of Labor Statistics, focused on the local area unemployment statistics, resulting in the Table 10.8, including the unemployment rate, as well as the ranking per American State, as follows.

As it can be seen in the Table 10.8, the American States that present the lowest unemployment rate in the country are North Dakota (1.9), South Dakota (2.1), Maryland (2.3), Vermont(2.3), Nebraska (2.5), Kansas (2.6), New Hampshire (2.6), Minnesota (2.7), Utah (2.8), and Wyoming (2.8); on the other hand, the States that have a higher level of unemployment rate are Nevada (5.3), California (5.2), District of Columbia (5), New Jersey (4.8), Illinois (4.7), Alaska (4.6), Washington (4.6), New York (4.5), Connecticut (4.4), Kentucky (4.3).

Using the information gathered for the present study, a Neural network analysis (nna) using multilayer perceptron network was performed in the data corresponding to every relevant variable in the study, which was used to categorize the probability of the decent work competitiveness regarding the textile industry per American State.

In that sense, the software "Statistical Package for Social Sciences" (SPSS) was used to perform the ANN on the dataset prepared for the research goals in the present work, which according to the official website, the program uses nonlinear data modeling to discover complex relationships among data, being helpful in the forecasting and decision-making process, based on multilayer perceptron networks that use feedforward architectures—data only moves from input nodes through the

[12] https://www.eia.gov/energyexplained/electricity/electricity-in-the-us-generation-capacity-and-sales.php.

Table 10.8 Unemployment rates for American States (seasonally adjusted)

State	Rate	Rank
Alabama	2.9	11
Alaska	4.6	45
Arizona	4.2	40
Arkansas	3.7	32
California	5.2	50
Colorado	3.4	24
Connecticut	4.4	43
Delaware	4.1	37
District of Columbia	5	49
Florida	3.1	16
Georgia	3.1	16
Hawaii	3.1	16
Idaho	3.3	21
Illinois	4.7	47
Indiana	3.5	27
Iowa	3	12
Kansas	2.6	6
Kentucky	4.3	41
Louisiana	4.1	37
Maine	3.4	24
Maryland	2.3	3
Massachusetts	3	12
Michigan	4	35
Minnesota	2.7	8
Mississippi	3.2	19
Missouri	3.3	21
Montana	3.3	21
Nebraska	2.5	5
Nevada	5.3	51
New Hampshire	2.6	6
New Jersey	4.8	48
New Mexico	4	35
New York	4.5	44
North Carolina	3.5	27
North Dakota	1.9	1
Ohio	3.7	32

(continued)

Table 10.8 (continued)

State	Rate	Rank
Oklahoma	3.5	27
Oregon	4.1	37
Pennsylvania	3.4	24
Rhode Island	3.6	31
South Carolina	3	12
South Dakota	2.1	2
Tennessee	3.5	27
Texas	3.9	34
Utah	2.8	9
Vermont	2.3	3
Virginia	3	12
Washington	4.6	45
West Virginia	4.3	41
Wisconsin	3.2	19
Wyoming	2.8	9

Source Own elaboration using databases from U.S. BUREAU OF LABOR STATISTICS[13] (n.d.)

hidden layer of nodes to output nodes[14]; applying the statistical test, the following results appeared.

Neural Network Analysis (Multilayer Perceptron Network)

As we know, Artificial Neural Networks (ANN), inspired by the structure of neurons in the human brain, have become popular supervised learning models and are increasingly prevalent in various applications; these models utilize a series of activation units, or neurons, to predict specific outcomes, and each neuron produces an output by applying a transformation function to its input.

In the use of the multilayer perceptron classifier, we employ the hyperbolic tangent function as the activation function, configure # hidden layers with ## neurons each, and utilize the lbfgs solver for weight optimization. This solver, part of the quasi-Newton methods family, offers faster convergence for small datasets compared to other optimizers. Neural Networks were chosen for their exceptional ability to recognize correlations and uncover hidden patterns in raw data, as well as their robust clustering and classification capabilities.

[13] https://www.bls.gov/web/laus/laumstrk.htm.

[14] https://www.ibm.com/products/spss-statistics/neural-networks.

Thus, for the present research, the MLP (multi-layer perceptron) neural network underwent training and evaluation using a dataset tailored to the research objective. The dataset was partitioned into training, validation, and testing subsets to ensure robust evaluation of the model's performance. During the training phase, the network's parameters were optimized via backpropagation and gradient descent algorithms to minimize the Cross Entropy Error. Subsequently, the trained model underwent evaluation on the testing subset to compute the Cross Entropy Error and Percent Incorrect Predictions metrics (Table 10.9).

To understand better the architecture of the model, the following Table 10.10 shows the network information, that is also being represented in terms of 5 factors or variables included in the present studio, 220 units of information, only 1 hidden layer consisting of 22 units, as follows:

Table 10.9 Case processing summary

		N	Percent
Sample	Training	132	0.729282
	Testing	49	0.270718
Valid		181	1
Excluded		7	–
Total		188	–

Source Own elaboration using the software SPSS (2024)

Table 10.10 Network information

Input Layer	Factors	1	Real gross domestic product (GDP) 1 (Millions of chained 2017 dollars)
		2	College completion rate per State
		3	Numbers of nonfatal occupational injuries and illnesses
		4	State renewable electricity profiles net generation (gigawatthours)
		5	Unemployment rates for states, seasonally adjusted
Number of units[a]			220
Number of hidden layers			1
Units in hidden layer 1[a]			22
Activation function			Hyperbolic tangent
Dependent variables			Competitiveness
Number of units			3 categories
Activation function			Softmax
Error function			Cross-entropy

[a] Excluding the bias unit
Source Own elaboration using the software SPSS (2024)

After processing the data, Table 10.11 shows the overall results, for the training, and testing samples, considering the 72–28% approximate partition of the data, as follows:

The following steps regards the results for classification, which includes the correct percentage in the classification process, where it is possible to observe that the comprehensive construct built to understand decent work competitiveness is strong and robust, and thus, a trustable model for studying and interpreting such complex concept in the American textile industry, as follows (Table 10.12).

Table 10.11 Model summary

Training	Cross entropy error	1.930187
	Percent incorrect predictions	0.007576
	Stopping rule used	1 consecutive step(s) with no decrease in error
	Training time	0:00:00.11
Testing	Cross entropy error	0.0008
	Percent incorrect predictions	0
Dependent variable: competitiveness		

Error computations are based on the testing sample
Source Own elaboration using the software SPSS (2024)

Table 10.12 Classification

Sample		Predicted			
		Competitive	Competitive advantage	Not competitive	Percent correct
Training	Competitive	67	0	0	1
	Competitive advantage	0	32	0	1
	Not competitive	1	0	32	0.969697
	Overall percent	0.515152	0.242424	0.242424	0.992424
Testing	Competitive	23	0	0	1
	Competitive advantage	0	14	0	1
	Not competitive	0	0	12	1
	Overall percent	0.469388	0.285714	0.244898	1
Dependent variable: competitiveness					

Source Own elaboration using the software SPSS (2024)

It is noticeable that the level of correct classification is over 96% in general, even higher than 99% for competitive advantage, which represents the overall goals of the present study.

Conclusion

This paper presents a rigorous analysis of the performance of a (MLP) neural network in the context of SDG 8: Decent work and economic growth, with a specific focus on the American textile industry. The study integrates a comprehensive set of variables crucial to SDG 8 objectives, including state-level GDP, educational attainment, occupational safety standards, clean energy adoption, and unemployment rates across American states.

The MLP neural network underwent meticulous training and evaluation using a dataset meticulously curated to capture the intricacies of the textile industry within the American context. Evaluation was conducted through two primary metrics: Cross Entropy Error (yielded a value of 1.930187) and Percent Incorrect Predictions (exhibited a value of 0.007576), highlighting the MLP's exceptional accuracy and reliability in forecasting outcomes pertinent to decent work and economic growth within the American textile industry, considering factors such as unemployment rates and workplace safety compliance.

In relation to the results, it is possible to argument that, to properly address decent work competitiveness, it is necessary to evaluated such complex concept in relation to how well it can handle working, economic, social and environmental challenges, including a proper analysis if the conditions for decent work, revealing a synergistic effect of a variety components in high-income countries.

References

Barford A, Coombe R, Proefke R (2021) Against the odds: young people's high aspirations and societal contributions amid a decent work shortage. Geoforum 121:162–172

Bianchi RV, de Man F (2021) Tourism, inclusive growth and decent work: a political economy critique. In justice and tourism, pp 220–238). Routledge

Chada L, Mashavira N, Mathibe MS (2022) The role of decent work in the Zimbabwean retail sector: testing a job engagement and turnover intention model. SA J Human Res Manage/SA Tydskrif vir Menslikehulpbronbestuur 20(0):a2029. https://doi.org/10.4102/sajhrm.v20i0.2029

Esparza S "WHY EVERY ORGANIZATION NEEDS AN AUGMENTED REALITY STRATEGY" (2018) Harvard Business Review, United Stated. Letter. https://www.researchgate.net/publication/351059991_Augmented_Reality_A_factor_to_change_peoples_values. Accessed 31 May 2024

Finatto CP, Aguiar Dutra AR, Gomes da Silva C, Nunes NA, Guerra JBSODA (2023) The role of universities in the inclusion of refugees in higher education and in society from the perspective of the SDGS. Int J Sustain High Educ 24(3):742–761

Frey DF (2018) Economic growth, full employment and decent work: the means and ends in SDG 8. In The sustainable development goals and human rights, pp 122–142. Routledge

Gusmão Caiado RG, Leal Filho W, Quelhas OLG, Luiz de Mattos Nascimento D, Ávila LV (2018) A literature-based review on potentials and constraints in the implementation of the sustainable development goals. J Clean Prod 198:1276–1288. ISSN 0959-6526

Kemechian T, Sigahi TF, Martins VW, Rampasso IS, de Moraes GHSM, Serafim MP, Anholon R (2023) Towards the SDGs for gender equality and decent work: investigating major challenges faced by Brazilian women in STEM careers with international experience. Discover Sustain 4(1):11

Leal Filho W, Azeiteiro U, Alves F, Pace P, Mifsud M, Brandli L, Caeiro SS, Disterheft A (2018) Reinvigorating the sustainable development research agenda: the role of the sustainable development goals (SDG). Int J Sustain Develop World Ecol 25(2):131–142. ISSN 1350-4509

Makarenko I, Plastun A, Petrushenko Y, Vorontsova A, Alwasiak S (2021) SDG 4 and SDG 8 in the knowledge economy: A meta-analysis in the context of post-COVID-19 recovery. Knowledge and Performance Management 5(1):50

Mariano KLP, Menezes RG, Franzoni M (2023) Regionalism and the Agenda 2030: inequality and decent work in Mexico. Rev Brasil Polít Int 66:e016

Nisa ZA, Schofield C, Neat FC (2022) Work below water: the role of scuba industry in realising sustainable development goals in small island developing states. Mar Policy 136:104918

Porter ME (1991) Towards a dynamic theory of strategy. Strateg Manag J 12(S2):95–117

Rai SM, Brown BD, Ruwanpura KN (2019) SDG 8: decent work and economic growth–A gendered analysis. World Dev 113:368–380

Ruxho F, Ladias CA, Tafarshiku A, Abazi E (2023) Regional employees perceptions on decent work and economic growth: labour market Of Albania And Kosovo. Regional Sci Inquiry 15(2):13–23

Sachs JD, Schmidt-Traub G, Mazzucato M et al. (2019) Six transformations to achieve the sustainable development goals. Nat Sustain 2:805–814. https://doi-org.pbidi.unam.mx:2443/10.1038/s41893-019-0352-9

Schroeder P, Anggraeni K, Weber U (2019) The relevance of circular economy practices to the sustainable development goals. J Ind Ecol 23(1):77–95

Sinha A, Sengupta T, Alvarado R (2020) Interplay between technological innovation and environmental quality: formulating the SDG policies for next 11 economies. J Clean Prod 242:118549

Wood SLR, Jones SK, Johnson JA, Brauman KA, Chaplin-Kramer R, Fremier A, Girvetz E, Gordon LJ, Kappel CV, Mandle L, Mulligan M, O'Farrell P, Smith WK, Willemen L, Zhang W, DeClerck FA (2018) Distilling the role of ecosystem services in the Sustainable Development Goals. Ecosyst Serv 29:70–82. https://doi.org/10.1016/j.ecoser.2017.10.010

Zehri C, Bendahmane MEA, Inaam Z, Sekrafi H (2024) Exploring the Nexus of decent work, financial inclusion, and economic growth: a study aligned with SDG 8. Sustainable Futures 100213

Chapter 11
Critical Evaluation of Successful Strategies for Sustainable Manufacturing in the Textile Sector and Their Impact on the Sustainable Development Goals: Case Studies in Colombia

Juan Angel Chica-Urzola

Introduction

The way we dress not only reflects our personality but also our economic, social, and political status, as well as our self-esteem. Fashion can be considered an art form that allows consumers to express themselves according to their own criteria and preferences, either by adopting a personal style or by following the latest trends (Nguyen et al. 2021), offer another definition of fashion, describing it as "an activity where individuals express personal tastes and interests through clothing".

The size of the textile market is estimated at $748 billion in 2024 and is expected to reach $889.24 billion in 2029, growing at a compound annual growth rate of 3.52% during the forecast period (2024–2029) (Yadav and Kamble 2024). The COVID-19 pandemic drastically challenged the textile industry in 2020. Asia, one of the largest markets for the global textile industry, suffered from prolonged lockdowns and restrictions in most Asian countries, along with a sudden drop in international demand for its products (Baliyan et al. 2024). The loss was particularly high in countries where the textile industry represented a larger proportion of exports. According to the International Labour Organization (ILO) study, global textile trade plummeted during the first half of 2020. Additionally, exports to major buying regions such as the European Union, the United States, and Japan fell by around 70%. The industry also faced several supply chain disruptions due to shortages of cotton and other raw materials (Yadav and Kamble 2024).

J. A. Chica-Urzola (✉)
Sustainable Development Industrial Engineering Program, Faculty of Science and Engineering, Universidad del Sinú, Córdoba, Colombia
e-mail: juanchica@unisinu.edu.co

A. E. Toscano-Hernández et al. (eds.), *Sustainability, Stakeholders and Marketing in the Textile Sector*, SDGs and Textiles, https://doi.org/10.1007/978-981-97-6952-0_11

The textile industry is a constantly growing market, with its main competitors being China, the European Union, the United States, and India. China is the world's leading producer and exporter of textiles and raw garments. The United States is the main producer and exporter of raw cotton, as well as the largest importer of textiles and raw garments. The European Union's textile industry includes Germany, Spain, France, Italy, and Portugal at the forefront, accounting for more than one-fifth of the global textile industry. India is the third-largest textile industry and is responsible for over 6% of total global textile production (Boschmeier et al. 2024). Rapid industrialisation in developed and developing countries and evolving technology are helping the textile industry to have modern facilities capable of producing high-efficiency fabrics. These factors are helping the textile industry to record more revenue during the study period and are expected to further assist the industry during the forecast period (Bradu et al. 2023).

Fast fashion refers to the market introduction of large volumes of clothing produced by the fashion industry in short periods, aimed at satisfying trends and an artificially created need for renewal (Granados and Chavez 2021). Reducing or even eliminating this continuous need for renewal is what makes a consumer sustainable, even more so than buying sustainable products. This fashion is characterised by its low durability due to the poor quality of the materials of the garments, reflected in their lower price, making them more attractive to consumers as "low-cost" items. (Nguyen et al. 2021). To meet the demand for fast fashion, these textiles are mainly produced in Southeast Asian countries under poor or non-existent working conditions. Bangladesh, China, Malaysia, Indonesia, and India are notable examples. Simply looking at the labels on the clothes we buy reveals this fact (Silván-Ferrero et al. 2023).

Sustainability in the textile sector has become a central concern in the global industry, driven by growing environmental and social awareness. In a world where fast fashion has dominated for decades, a focus on more responsible and ethical practices is gaining ground (Medina et al. 2019).

One of the key areas of focus is reducing the environmental impact of textile production. Companies are adopting cleaner manufacturing practices, using technologies and processes that minimise water, energy, and chemical consumption (da Silva et al. 2021). In addition, the use of sustainable and recycled materials in garment manufacturing is being encouraged, reducing dependence on non-renewable resources.

Sustainability also encompasses social and ethical aspects throughout the supply chain. Companies are working to improve working conditions in textile factories, ensuring fair wages and safe conditions for workers. In addition, initiatives are being implemented to combat child labour and promote gender equality in the sector (Nguyen et al. 2023).

Circular fashion is also gaining momentum as a key approach to improving sustainability in the textile industry (Papamichael et al. 2022). This involves designing garments with materials that can be recycled or reused at the end of their useful life, thus creating a continuous life cycle for textile products.

Consumers are also playing a crucial role in promoting sustainability in the textile sector (Zaman and Kusi-Sarpong 2024). More and more people are opting for brands that prioritise sustainability and are willing to pay a fair price for responsibly produced garments. Education and awareness of sustainable practices is also increasing, leading to a greater demand for transparency and ethics throughout the supply chain (Abbate et al. 2024).

Theorical Background

Sustainable Manufacturing

In relation to sustainable development, Baumgartner and Ebner (2010) pick up on Griggs, Stafford-Smith, Gaffney, Rockström, Öhman, Shyamsundar and Noble in 2013, when they state that it represents rising social expectations for business conduct (Baumgartner and Ebner 2010). A perspective shared by the United Nations when presenting the Sustainable Development Goals (SDGs) as these are planned to act as framework conditions for the global economy in the coming years (Vildåsen et al. 2017).

The industrial sector is directly or indirectly responsible for environmental degradation (Jiang et al. 2018), or at least a significant part of it. In response to this situation, many manufacturers, both domestically and globally, are exploring ways to reduce the effect of their industrial activities on the environment and society, and thus improve their sustainability performance.

Many essential human needs can only be satisfied through goods and services provided by industry. The products of industry form the material basis of contemporary living standards. And Colombia is no stranger to this, as all nations rightly require and aspire to efficient industrial bases to meet changing needs (Brundtland 1987).

Approximately 2,000,000 years ago, with the appearance of Homo habilis, the use of natural elements as satisfiers of human needs, both individual and collective, began to intensify. Already later, the Greeks observed this phenomenon of needs and their satisfaction, whether from the hedonic or eudaimonic perspective, as a form of important interaction of man with the environment materialised through the manufacture and use of the product, be it a good or a service, as a fundamental satisfier or vehicle to achieve the satisfaction of his different needs, even giving uses or attributing characteristics at different levels to these (Chica-Urzola and Serna Mendoza 2018).

As the world's population grew, their needs changed and increased; this in turn led to changes in the way of generating satisfactions and making use of the resources of the territory and the planet for this purpose; This race, as recognised in the Brundtland Report (Brundtland 1987), requires the permanent use of raw materials, constant increases in productivity and the generation of material goods in large quantities, which has imposed a very high economic cost, as well as a heavy burden of environmental impacts (Brundtland 1987).

When, in 1972, the Meadows Report, named Limits to Growth, was published, the need for a change in the model of resource consumption began to be considered, since the linear model, based on inputs, transformation and outputs, is only sustainable in the face of the infinite existence of resources. He does this in the face of what he calls the imminence of resource depletion if the prevailing model is not changed and the halt in industrial growth because of this (Meadows 1972).

This sentence set off alarm bells in the different sectors involved in and responsible for the production and consumption of the planet's resources. Especially when reflecting on the role of industry, which, as recognised in the Brundtland Report (1987), "is fundamental to the economies of modern societies and an indispensable engine of growth" (Brundtland 1987, p. 173). And even more so when, within this reflection, aspects such as the following are considered: The nature of companies makes them those elements that take natural resources and transform them into finished products, therefore, they are seen as the predators of the same (Hu et al. 2017). During the transformation process of these raw materials and incoming inputs, the company generates by-products and emissions both liquid and solid, fumes, vapours, and aerosols whose impact on the environment is polluting (Silva and Lopes 2017; Sharma et al. 2019). Once the cycle of product consumption by the user is over, the product is disposed of inappropriately, causing stress at the disposal sites as the load received is often greater than the absorption rate of the planet, both in quantity and time (Zhaowen Cheng et al. 2019; Agrawal and Singh 2019). Companies have been seen as places where economic growth is privileged to the detriment of worker and environmental well-being, often without applying compensation strategies to the territory or generating negative impacts on it (Ciplet 2021; Pagell and Shevchenko 2014). And, finally, lack of knowledge, confusion or misunderstanding of what is meant by Corporate Sustainability or Sustainable Business Development. This is largely due to the enormous number of concepts and interpretations that have been made about sustainable development since this concept began to be used in the world. On the other hand, because of the industrial sector's disinterest in getting into this area (Fayomi et al. 2019).

The adoption of the Brundtland Report by international organisations has had consequences for the global industrial sector and for the Colombian textile sector. Some of these have been identified by several authors, highlighting those associated with regulatory and legislative pressures on the textile sector industries, predominant changes in the pattern of consumer demand towards more sustainable products and practices (Saha et al. 2022; Gbededo et al. 2018). Therefore, like industrial organisations, companies in the textile sector now suffer not only from internal pressure to achieve economic benefits, but also from external pressures to establish the system of environmental and social responsibility (Jiang et al. 2018).

Along the same lines, some authors argue that there is an apparent agreement in the literature regarding the existence of pressures that the organisational environment is exerting on companies to adopt a series of solutions aimed at protecting and conserving the natural environment, either through existing legislation or through market mechanisms, which is why it is increasingly common to find companies that present themselves as socially responsible and environmentally aware of the impact that their activity has on the natural environment (Jayaraman et al. 2023; Baah et al. 2021; Accinelli and De la Fuente-García 2013). A situation from which the manufacturing sector that is the subject of this research does not escape.

As a result, the current global focus is now on supporting and forcing industries to implement cleaner and more efficient production practices that enable the development of products and services with reduced negative environmental and social impacts (Gbededo et al. 2018). As rightly expressed by Gbolarumi, Wong & Olohunde (2021), or Rosen and Kishawy (2012), industries, and in general the entire textile sector, are faced, in addition to economic efficiency, with evaluating environmental and social objectives and impacts in the advancement of manufacturing operations, technologies and competitive position (Gbolarumi et al. 2021; Gbededo et al. 2018).

Complementarily, as Chang and Cheng (2019) rightly state, multiple authors have recognised that sustainable development provides competitive advantages over competitors. Others state that the benefits of sustainable development efforts are reflected not only in quantifiable financial performance and other economic indicators, but also in many managerial ways. For its part, sustainable development enhances corporate reputation, obtains legality of management decisions, promotes labour relations, attracts resources, and reduces stakeholder pressure on the company (Chang and Cheng 2019).

To unify, or at least reduce the number of variant concepts of sustainable manufacturing, the US Department of Commerce, in 2009, defines sustainable manufacturing as "the creation of manufactured products that use processes that minimise negative environmental impacts, conserve energy and natural resources, are safe for employees, communities, and consumers, and are economically sound" (Roberts and Ball 2014, p. 161). But despite this effort, their acceptance has not been universal and, as a result, the proliferation of more and new definitions has continued.

It is from the different business visions that sustainable development can have, so that corporate sustainability performance has come to be defined in many ways but, if there is one thing, they all converge on, it is to approach it as a strategy for companies to seek harmony between economic profit, environmental and social responsibility, and other stakeholders (Jiang et al. 2018). From this convergence on how corporate sustainable development can be defined, the main forms of corporate operationalisation of this conceptualisation emerge (Moldavska and Welo 2017): Manufacturing for sustainability, which refers to the development of products to help society to be sustainable, and sustainability of the manufacturing, which refers to the production of products through a sustainable production system.

However, as Henao et al. (2019) rightly say, this is easier said than done. Despite the wake-up call generated by the 1987 "Brundtland Report" of the United Nations World Commission on Environment and Development, evidence suggests that many companies continue to relegate environmental protection and social responsibility to places below or subordinate to economic performance (Henao et al. 2019).

In the same vein, it is also suggested that research evidence on these issues suggests that there is a conflict between corporate objectives and the attractiveness of sustainable development. It seems that good environmental performance or socially responsible behaviour towards the natural environment and society on the part of companies conflicts with the central objective of any company: profit maximisation (Camilleri, 2022; Accinelli Gamba and De la Fuente-García 2013).

As Morioka and de Carvalho (2016) put it, by orienting the company's activities towards sustainable development, the organisation matches its responsibilities to society and the environment at the institutional, organisational and individual levels (Morioka and de Carvalho 2016).

Sustainability and the Textile Industry

The textile industry is one of the largest in the world and, unfortunately, also one of the most unsustainable in environmental, social and economic terms (Xu et al. 2018). Over the past decades, the sector has experienced exponential growth, largely driven by fast fashion and rampant consumption (Pastor Tomas, 2023). However, this growth has been accompanied by several serious problems that threaten the health of the planet and human communities (Castro 2021).

One of the main concerns is the environmental impact of the textile industry. From the production of raw materials to the manufacturing process and the transportation of finished products, each stage of the supply chain has a significant impact on the environment (Islam et al. 2022). For example, conventional cotton production requires large amounts of water and pesticides, which depletes water resources and pollutes the soil and air (Islam et al. 2022). In addition, many of the chemicals used in the dyeing and finishing processes are highly toxic and can cause irreparable damage to local ecosystems (Pastor Tomas, 2023).

Another major problem is the waste generated by the textile industry. Fast fashion is based on accelerated production cycles and constant inventory turnover, leading to overproduction and an increase in the amount of textile waste (Argüello Gavilánez, 2022; Fernández-López et al. 2020). It is estimated that millions of tonnes of clothing are produced each year and end up in landfills or incinerators, further contributing to the global waste crisis (Argüello Gavilánez 2022).

Another form of waste, generated by the textile industry, is associated with the obsolescence of its products. Planned obsolescence and perceived obsolescence are phenomena that have permeated this industry and have had a significant impact on consumption patterns and sustainability (Bustos Mundaca 2022). Experiences of entrepreneurship in slow fashion: alternative proposals moving between rationalities... These concepts are closely related and have important implications for the way garments are designed, produced and marketed (Correa Morales 2023).

Planned obsolescence refers to the practice of designing products with a limited useful life, with the aim of encouraging repeat purchases and maintaining high levels of consumption (Antonucci 2021). In the textile industry, this can manifest itself in several ways. For example, garments may be designed with low-quality materials or weak seams that wear out quickly, forcing consumers to replace them frequently (Cortés Hernández et al. 2023; Antonucci 2021).

In addition, planned obsolescence can also manifest itself through changing fashions and ephemeral trends (Antonucci 2021; Castro 2021). Fast fashion brands tend to produce increasingly frequent fashion collections at low prices, leading to a constant cycle of buying and discarding clothes that are in fashion for only a short period of time (Vinlove 2021).

Perceived obsolescence, on the other hand, refers to the idea that products become obsolete not because they cease to function properly, but because they are perceived as outdated or outmoded (Segurajáuregui Álvarez 2016). In the textile industry, this is exacerbated by rapid inventory turnover and the constant introduction of new fashion collections (Carrera Gallissà 2017). Consumers may feel pressured to follow the latest trends and discard perfectly functional garments simply because they are no longer fashionable (Castro 2021).

The combination of planned obsolescence and perceived obsolescence in the textile industry has led to an increase in clothing waste and a significant environmental impact (Carrera Gallissà 2017). It is estimated that millions of tonnes of clothing end up in landfills or incinerators each year, contributing to the global waste crisis and environmental degradation (Cortés Hernández et al. 2023).

In addition, rapid inventory turnover and mass production of garments also have negative implications for labour conditions in the textile supply chain (Castro 2021). Fast fashion brands often resort to low-cost factories in developing countries, where workers face poor working conditions, low wages and lack of basic labour rights (Choi 2023).

In addition to environmental impacts, the textile industry also faces serious social and economic problems. In many developing countries, where much of the textile production is concentrated, workers face hazardous working conditions, low wages, and lack of basic labour rights (Quintero Arango et al. 2019). Child labour and labour exploitation are unfortunately common in many textile factories, where workers are forced to work long hours for minimal wages and in unsafe conditions (Martinez 2021; Norando 2020).

Unsustainability in the textile sector also has global economic ramifications. Fierce competition and the constant search for lower costs have led to the offshoring of production to countries with cheap labour and lax regulations (Sheffi, 2023). This has had a devastating impact on local economies in developed countries, where closed factories have left thousands of people unemployed and entire communities in crisis (Vinlove 2021).

Despite these challenges, there are reasons to be optimistic about the future of the textile industry. More and more companies are recognising the need to adopt more sustainable and ethical practices in their operations (Correa Morales 2023; Jiménez Castellón 2023; Medina Fábrega 2023). From luxury fashion brands to small independent companies, there is a growing interest in producing garments more responsibly, using sustainable materials, reducing waste and improving labour conditions throughout the supply chain (Islam et al., 2022).

Consumers are also playing a key role in promoting sustainability in the textile sector (Pinilla & González, 2014). More and more people are opting for brands that prioritise transparency and ethics and are willing to pay a fair price for responsibly produced products (Rodríguez Bugada, 2022; Pitre-Redondo et al. 2020; Gómez et al. 2009). The demand for sustainable fashion is growing rapidly, which is putting pressure on companies to change their practices and adopt more responsible approaches (Cubero 2021).

To address these issues, a fundamental change is needed in the way garments are designed, produced and consumed (Desore and Narula 2018). This includes, as Carrera Gallissa (2017) puts it, a greater emphasis on product quality and durability, as well as the adoption of more sustainable production practices.

Sustainable Development Goals (SDGs)

The Sustainable Development Goals (SDGs) are a set of 17 global targets set by the United Nations in 2015 as part of the 2030 Agenda for Sustainable Development. These goals represent an urgent call to action to address the most pressing challenges facing our world, from eradicating poverty to protecting the environment and promoting peace and justice (UN Assembly 2015).

A distinctive feature of the SDGs is their integrated and holistic approach (Pradhan et al. 2017). They recognise that global challenges are interconnected and that effective solutions require close and coordinated collaboration between governments, civil society, the private sector and citizens (Moldavska and Welo 2017). Furthermore, the SDGs consider the economic, social and environmental dimensions of sustainable development, recognising the need for a balanced and long-term approach to achieve a prosperous and sustainable future for all (Leal Filho et al. 2018).

The SDGs represent a powerful framework for global action and mobilisation. Since their adoption, they have inspired a wide range of initiatives and policies around the world, from community development programmes to national sustainable development strategies. They have also provided a common language and a set of shared goals that unite different actors in the quest for a more just, equitable and sustainable world (Pradhan et al. 2017).

Despite progress on many fronts, many challenges remain to be addressed to achieve the SDGs by 2030. The COVID-19 pandemic has exacerbated many of the existing gaps in areas such as poverty, health, education, and gender equality, underlining the urgency of redoubling efforts to achieve significant progress in the coming years (Clemente-Suárez et al. 2022).

In the framework of sustainable development, the role of business is fundamental and indisputable, and this includes the national manufacturing sector, in this sense the United Nations Development Programme—UNDP—recognises that for the achievement of the Sustainable Development Goals—SDGs—the private sector, which includes businesses of all sizes, types of economy and operational scope, are the engine of inclusive economic growth and job creation (UNDP 2016). Manufacturing firms in Colombia, along with the global business sector, are expected to use their creativity and innovation to solve sustainable development challenges (Desore and Narula 2018). In addition, they are encouraged to adopt socially responsible practices related to decent employment, women's empowerment, investment in research and development, environmental protection, transparency and accountability that go beyond the purely economic sphere (Sanchez-Planelles et al. 2022).

The relationship between the textile industry and the Sustainable Development Goals (SDGs) is profound and complex, as the textile industry influences multiple aspects of human life and the environment (Gabriel and Luque 2020). The SDGs, established by the United Nations, are a global call to action to address the world's most pressing challenges, from poverty eradication to environmental protection (Akter et al. 2022; de Oliveira Neto et al. 2019).

One of the SDGs most directly related to the textile industry is SDG 12: Responsible Consumption and Production. Fast fashion and rampant consumerism have led to a culture of waste and overexploitation of resources (Bick et al. 2018). Excessive garment production, combined with short fashion lifecycles, has contributed to enormous textile waste generation and environmental degradation (Gabriel and Luque 2020). To address this problem, it is crucial that the textile industry adopts more sustainable production practices, reducing resource consumption, minimising waste and encouraging the reuse and recycling of materials (Niinimäki et al. 2020).

SDG 8: Decent work and economic growth, also has a direct connection to the textile industry. Garment production employs millions of people worldwide, especially in developing countries (de Oliveira Neto et al. 2019). However, the industry is also known for its precarious working conditions, low wages, safe and fair working conditions, and gender inequality at all levels of the supply chain (Tabales et al. 2021; Herrera Almanza and Corona 2020).

The textile industry plays a crucial role in achieving Sustainable Development Goal (SDG) number 9: "Industry, Innovation and Infrastructure". This sector is a driving force for technological innovation and job creation. However, it also faces significant challenges in terms of environmental sustainability and fair labour conditions. The adoption of clean and efficient technologies in textile production is essential to reduce its negative impact on the environment (Edreira Rico 2022). In addition, improving infrastructure in the areas where textile mills operate can help increase efficiency and reduce costs. Promoting research and development in sustainable textiles and fostering collaboration between companies, governments and civil society are essential steps to move towards a more equitable, innovative and sustainable textile sector that contributes to the achievement of the SDGs (Herrera Almanza and Corona 2020; Kynčlová et al. 2020).

SDG 6: Clean water and sanitation are another goal relevant to the textile industry, as garment production requires large amounts of water, especially in dyeing and finishing processes. Many textile factories are in regions where access to clean water is already limited, further exacerbating water resource scarcity and water pollution (Cai and Choi 2020). To contribute to SDG 6, the textile industry must adopt more water-efficient technologies and production practices, reduce pollution from industrial discharges, and work on the conservation and sustainable management of water resources (de Oliveira Neto et al. 2019).

In addition to these direct goals, the textile industry is also linked to other SDGs indirectly. For example, SDG 1: End Poverty, relates to the textile industry through its impact on employment and livelihoods of local communities. Similarly, SDG 13: Climate Action, is affected by the greenhouse gas emissions and energy consumption associated with garment production and transportation (Cai and Choi 2020).

Despite the challenges it faces, the textile industry also presents significant opportunities to contribute to the achievement of the SDGs. Many textile companies are adopting more sustainable approaches in their operations, investing in clean technologies, recycled materials and ethical production practices (Rathore 2022). Innovation in sustainable fashion is on the rise, with the development of new, more environmentally friendly fibers and materials, as well as business models focused on circularity and garment reuse (Thakker and Sun 2023).

The Textile Industry in Colombia

The textile industry in Colombia has been a crucial component of its economy and culture throughout the country's history. From its roots in artisanal production to modernization and globalization, the textile sector has undergone a series of transformations that have had a significant impact on Colombia's economy, society, and environment (Babativa et al. 2023).

The history of the textile sector in Colombia dates to pre-Columbian times, when indigenous communities wove fabrics with natural fibers such as cotton and alpaca wool (Babativa et al. 2023; Linares et al. 2008). This ancestral art was kept alive during the colonial era, when Spanish colonizers introduced European weaving techniques and organized textile production in workshops and factories (Harsanto et al. 2023).

During the nineteenth century, the Colombian textile industry experienced significant growth thanks to the production of cotton, wool and silk fabrics for the local market. However, it was in the twentieth century when the textile sector experienced a true revolution, with the introduction of modern machinery and the expansion of the industry towards the manufacture of clothing and technical textiles (Harsanto et al. 2023; Maldonado-Pinto & Portilla-Barco 2020; De Sociedades 2017; Santiago 1982).

The boom of the Colombian textile industry in the twentieth century had a remarkable economic impact, contributing significantly to GDP growth and employment generation in the country (González et al. 2018). The creation of textile factories and workshops in cities such as Medellín, Bogotá and Cali provided job opportunities for thousands of people, especially women, who played a key role in textile production (Andrade-Pérez et al. 2018).

However, the Colombian textile sector has also faced a series of challenges throughout its history. Competition from imported textile products, especially from China and other Asian countries, has affected the competitiveness of the national industry (Jarpa & Halog 2021; Castell et al. 2016). In addition, the lack of investment in technology and informality in the sector have limited its capacity for innovation and growth (Caicedo Manrique 2021).

In social terms, the textile sector has had a significant impact on the lives of Colombian communities, especially in rural areas where the production of natural fibers such as cotton and wool are concentrated. Job creation in the textile industry has provided a source of income for many families and has contributed to the economic development of these regions (Harsanto et al. 2023; Salgado Abril and Suárez Cordero 2016).

However, work in the textile sector has also been marked by precarious working conditions on some occasions, with low wages, long working hours and lack of job security (Harsanto et al. 2023). Despite efforts to improve working conditions, challenges persist in terms of labor rights and social protection for textile workers in Colombia (Mazo et al. 2018, 2020).

In terms of environmental impact, the textile industry is known for its high ecological footprint due to the intensive use of natural resources and the generation of waste and pollutants. The production of conventional textiles involves the use of large amounts of water, energy, and chemicals, which contributes to resource scarcity and air and water pollution (Harsanto et al. 2023; Pitre-Redondo et al. 2020).

Materials and Methods

The research developed was framed as interpretive type research, specifically critical evaluation. As Rodriguez (2018) mentions, the interpretive paradigm in research is a way of understanding scientific knowledge and reality. That is to say that this research model is based on the deep understanding of reality and the causes that have led it to be so. This scientific model is part of qualitative research, which seeks to study a topic in depth to fully understand it (Escobar et al.2018), but without neglecting the fundamental contribution of quantitative research, developed through a descriptive, explanatory type study, whose main sources of quantitative information, were obtained through statistics provided the sector and the in-depth review of related literature.

In addition to this, the approach to sustainability is based on a systemic approach, as a way of approaching and representing reality and at the same time, as an orientation towards a practice that stimulates thinking in interdisciplinary ways of working. A characteristic of this approach is its holistic and integrative perspective, where the importance of interrelationships and results derived from them is highlighted.

The application of the systemic vision to the Colombian textile industry is based on the multidimensional approach to sustainable development, connecting it at different levels of the system: economic, social, environmental, and institutional by studying the structure, environment and relationships immersed in it. Or as Garza (2007) maintains, when he conceptualizes sustainable development, presenting it as a frontier field of knowledge that integrates economic development and equity, environment and biodiversity, and culture and society (Garza 2007).

The procedure was developed in phases, because the systematicity in the execution of the activities allows the results of the preceding tasks to be used as inputs for the following tasks. This characteristic allowed the identification of three (3) stages to be developed, each of them related to a specific objective and described below.

PHASE 1: This phase is framed by the development of a descriptive type of research in which it is intended to characterize, with the use of statistical tools, the textile manufacturing sector in Colombia. The behavior of this scenario is determined by analyzing information from sources associated with the economic, environmental, and social dimensions of sustainable development and the manufacturing sector. Using cross analysis of data and information, the different states of the determinants of the scenario characterized in the framework of sustainable business development are established. As a result of this cross analysis, common characteristics, sector behaviors, transversal phenomena (both in time and in subsectors) that shaped the sector dynamics will be identified. This resulting information is presented in the form of diagrams and tables for better appreciation. The final purpose of this stage is to establish the framework, conditions, structure and dynamics of the scenario in which the textile manufacturing sector developed within the country.

PHASE 2: This second phase, characterized by developing an inquiry, as the selected type of research, in which the companies that have developed strategies that are framed in the context of sustainable manufacturing, in any of its two manifestations, are identified, and their impacts on the different SDGs, are identified according to the respective indicators.

PHASE 3: In this phase, use is made of the evaluative type of research. A combined analysis of the results of the previous phases will be carried out using content analysis, exploratory analysis and finally, a critical evaluation of the sustainable manufacturing strategies implemented by companies in the textile manufacturing sector in Colombia and their impact on the SDGs. As a result of a relational analysis, they are combined through a joint analysis to perform a critical evaluation. Finally, to issue a concept on the scenario of the Colombian textile manufacturing sector in the framework of sustainable manufacturing and the SDGs.

Results

As a result of the first phase, it is observed that the Colombian textile industry is one of the most prominent in Latin America, in this South American country, about 6,500 companies generate about one million jobs, of which about 600,000 are direct and the remaining estimated, indirect; being Bogota and Medellin the cities that participate in most of the textile production. Colombia has taken the number one position in exports of South American flat fabrics, in manufactures such as jeans, bras, girdles and control garments.

According to the Observatory Report of the Institute for Exports and Fashion (Inexmoda), In 2021, sales of textiles and apparel stood at 27.7 trillion pesos, in 2022 it was 30.30 trillion pesos, presenting a variation of 9.4% with respect to the

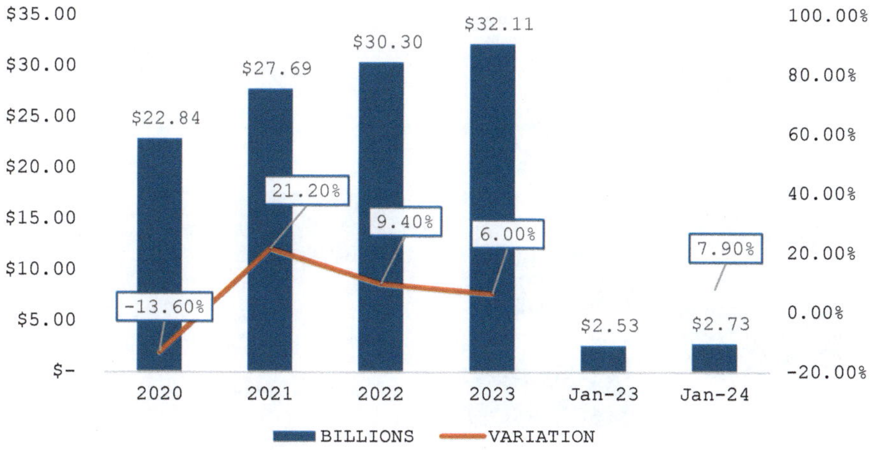

Fig. 11.1 Spending of Colombian households on textile sector products, between 2020–2023 and January 2024. *Source* Adapted from Inexmoda (2024)

immediately preceding year. Between January and December 2023, it was 32.11 trillion pesos, 6.0% more than the same period of 2022 and this industry participated with 9.4% of the total Gross Domestic Product (GDP). Figure 11.1 shows how Colombian households will spend on products in this sector from 2020 to 2023 and the first month of 2024.

Consumption month by month for the period between 2020 and January 2024, from here three (3) large groups can be observed according to the behavior presented: the group of the months January and February, which presents similar behaviors among them; the group of the months March-June and August-November, characterized by similar behaviors among them; and the group June and December, periods in which there is an increase in the consumption of textile sector products.

From the perspective of the value chain (see Fig. 11.2), the textile sector in Colombia involves different links ranging from the production of raw materials to the marketing of final products.

Production of Raw Materials: Colombia has production of some textile raw materials, such as cotton, wool, and synthetic fibers. However, domestic production of these raw materials is not always sufficient to meet demand, so some of the inputs must be imported. These imports may be legal or illegal (contraband).

Textile Manufacturing: There are numerous companies engaged in textile manufacturing in Colombia. These companies weave, dye, print, and manufacture fabrics for subsequent use in the production of clothing and other textile products.

Apparel Manufacturing: This is an important sector within the Colombian textile industry. There are many factories and garment workshops that produce a wide variety of garments, from casual wear to uniforms and work clothes.

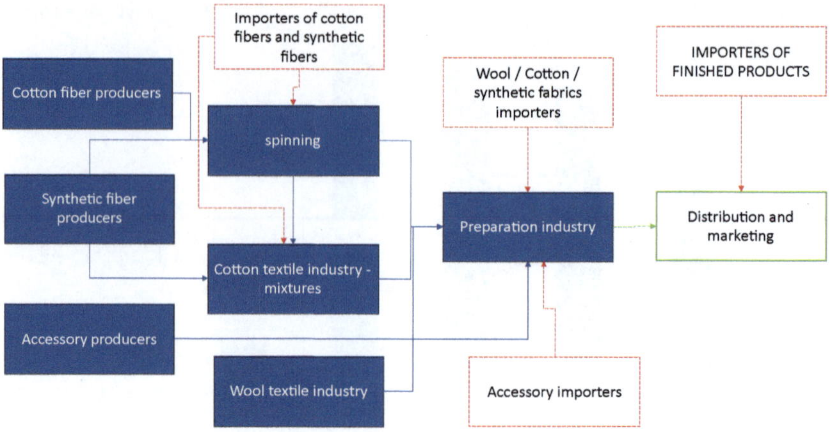

Fig. 11.2 Structure of the textile chain in Colombia. *Source* Adapted Legis (2012)

Fashion Design: Colombia also has a vibrant and creative fashion design sector. Many Colombian designers have gained international recognition for their innovative and high-quality fashion collections.

Marketing and Distribution: Colombian apparel and other textile products are marketed both domestically and internationally. There are a variety of distribution channels, including retail stores, department stores, boutiques, online marketplaces, and direct exports.

Regulation and Standards: Colombia's textile sector is subject to government regulations and standards ranging from quality and safety standards to trade and export policies. And, in the same field, associations and guilds: there are several associations and guilds in Colombia that represent the interests of textile companies and professionals. These organizations play an important role in promoting the industry, defending its interests, and providing support and resources to their members.

Now, going into the details of the textile manufacturing sector in Colombia, the production and sale of textiles closed 2023 with negative figures, being one of the industrial activities affected by lower consumption and the fall in orders from other countries (Inexmoda 2024). Spinning, weaving, and cotton were among the inputs with the largest declines. As the economic and financial variables become more flexible, the situation described above is illustrated in Fig. 11.3.

To better understand this behavior, the main structural characteristics of the textile manufacturing sector are described below.

Production of Fabrics and Accessories: In this stage, the fabrics and/or accessories that will serve as raw material for the manufacture of garments and other textile products are produced. These fabrics and/or accessories can be produced either in Colombia or abroad and then imported for use in the next stage of the process.

Dyeing and Printing: After the production of fabrics, some companies specialize in dyeing and printing them. This involves the application of dyes and designs on the fabrics to create different styles and finishes.

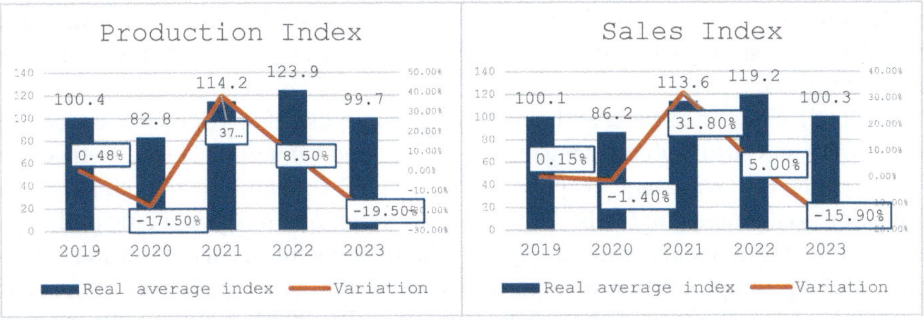

Fig. 11.3 Index of production and sales of textile products, year 2023. *Source* Data from the National Administrative Department of Statistics-DANE, 2024

Garment Manufacturing: This is one of the most important stages in the textile sector. In Colombia, there are numerous factories and garment workshops that transform fabrics into finished garments.

Subcontracting and Maquila: Many apparel companies in Colombia also perform subcontracting or maquila for international brands. This involves the manufacture of garments according to the designs and specifications of foreign brands, which contributes significantly to the Colombian economy and employment in the sector.

Product Design and Development: In addition to the physical production of apparel, Colombia also has a sector dedicated to the design and development of textile products. This is where designers create new styles, patterns and collections that are then produced in factories.

Exports and Imports: The Colombian textile sector is integrated into the global economy through the export and import of textile and apparel products. Colombia exports a variety of textile products to countries around the world, while importing raw materials and finished products to meet domestic demand.

To observe this structure more clearly, Fig. 11.4 is presented below.

Now, continuing with the methodological development of the research, we approach the Sustainable Development Goals (SDGs). The SDGs have a set of specific targets integrated into each of the 17 goals. Specifically, a selection was

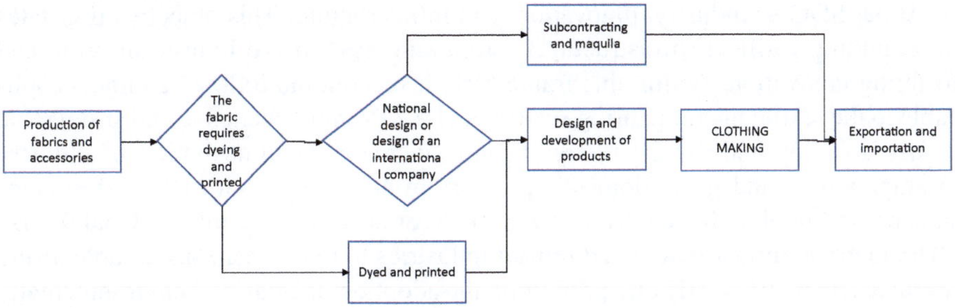

Fig. 11.4 Structure of the textile manufacturing sector in Colombia

made of those goals associated with goals 6, 7, 8, 8, 9 and 12, on which sustainable manufacturing strategies used by companies in the textile manufacturing sector can have an impact.

About SDG 6: Clean water and sanitation. The purpose of this SDG is to ensure the availability and sustainable management of water and sanitation for all. To this end, it sets out the following targets applicable to the textile manufacturing industry. The following goals stand out: 6.3 By 2030, improve water quality by reducing pollution, eliminating dumping, and minimizing the release of chemicals and hazardous materials, halving the proportion of untreated wastewater, and significantly increasing recycling and safe reuse globally. And goal 6.4 By 2030, significantly increase the efficient use of water resources in all sectors and ensure the sustainability of freshwater withdrawals and supplies to address water scarcity and reduce water stress.

In the case of SDG 7: Affordable and clean energy. This aims to ensure access to affordable, secure, sustainable, and modern energy. Targets associated with this goal, and which may have an impact on the textile sector include goal 7.2 By 2030, significantly increase the share of renewable energy in the overall energy mix. Goal 7.3 By 2030, double the global rate of improvement in energy efficiency. And goal 7.a By 2030, increase international cooperation to facilitate access to clean energy research and technology, including renewables, energy efficiency, advanced and cleaner fossil fuel technologies, and promote investment in energy infrastructure and clean technologies.

In reference to SDG 8: Decent work and economic growth. This SDG aims to promote inclusive and sustainable economic growth, employment, and decent work for all. To this end, it presents the following targets applicable to the textile manufacturing sector. It is in this framework that the goals 8.2 Achieve higher levels of economic productivity through diversification, technological upgrading, and innovation, including through a focus on labour-intensive, high value-added sectors. Goal 8.3 Promote development-oriented policies that support productive activities, decent job creation, entrepreneurship, creativity, and innovation, and encourage the formalisation and growth of micro, small and medium-sized enterprises, including through access to financial services. And goal 8.8 Protect labour rights and promote a safe and secure working environment for all workers, including migrant workers, in particular migrant women and those in precarious employment, are extracted.

About SDG 9: Industry, innovation, and infrastructure. This objective translates into building resilient infrastructures, promoting sustainable industrialisation and fostering innovation. Within this framework, it sets out the following targets applicable to the textile manufacturing sector: goal 9.2 Promote inclusive and sustainable industrialisation and, by 2030, significantly increase the contribution of industry to employment and gross domestic product, in accordance with national circumstances, and double this contribution in the least developed countries. Goal 9.4 By 2030, upgrade infrastructure and retrofit industries to make them sustainable, using resources more efficiently and promoting the adoption of clean and environmentally sound technologies and industrial processes, with all countries acting in accordance with their respective capacities. And goal 9.5 Enhance scientific research and improve technological capabilities in industrial sectors in all countries, in particular developing countries, including by promoting innovation and significantly increasing the number of research and development personnel per million inhabitants and public and private sector expenditure on research and development by 2030.

And finally, in relation to SDG 12: Responsible production and consumption: This goal aims to ensure sustainable consumption and production patterns. It sets out the following targets: goal 12.2 By 2030, achieve the sustainable management and efficient use of natural resources. Goal 12.4 By 2020, achieve the environmentally sound management of chemicals and all wastes throughout their life cycle, in accordance with agreed international frameworks, and significantly reduce their release to air, water and land to minimise their adverse effects on human health and the environment. And goal 12.5 By 2030, significantly reduce waste generation through waste prevention, reduction, recycling, and reuse activities. Figure 11.5 shows every selected SDG with its own goals applied to the textile industry.

Figure 11.6 shows in more detail in which SDGs are reflected the actions framed in the field of sustainable manufacturing, in either of its two (2) expressions, namely, manufacturing for sustainability and manufacturing sustainability, which are implemented in the textile manufacturing sector industries.

Taking the above as a framework for the relationship between the SDGs and sustainable manufacturing strategies, the second phase of the study begins, according to the methodology proposed, as a result it was discovered that the strategies implemented by the industries in the textile manufacturing sector, in the context of sustainable manufacturing, stand out more for its impact on SDGs, are:

Strategies implemented by Textiles Fabricato—Tejicondor: Fábrica de Hilados y Tejidos del Hato (Fabricato) was founded in 1920 and started operations in 1923, while Textiles El Cóndor (Tejicóndor) was founded in 1934. These companies were

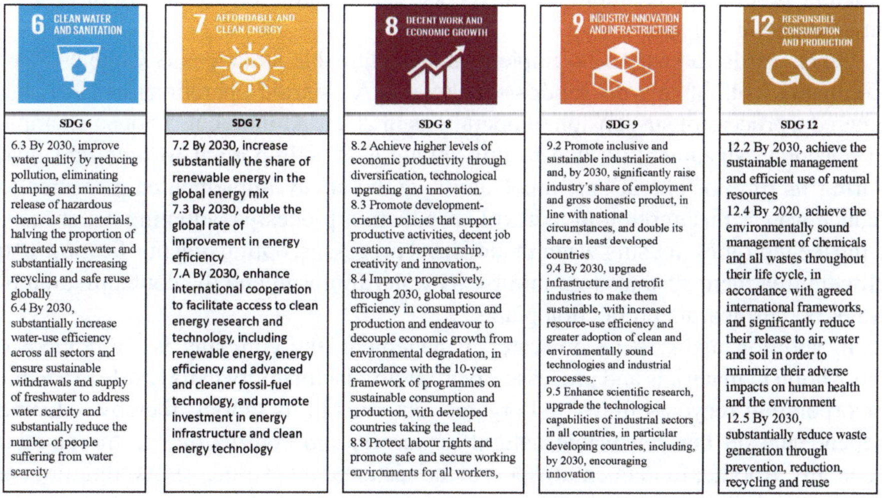

Fig. 11.5 Shows the different goals identified inside of SDGs

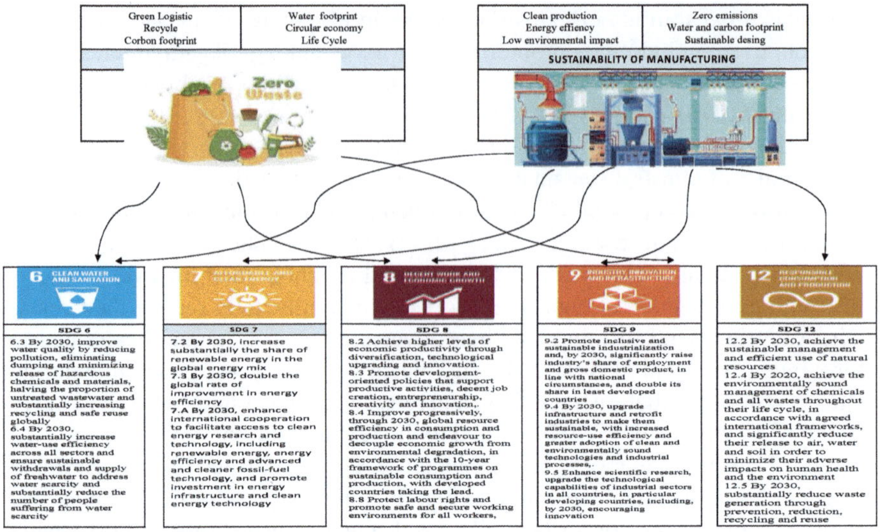

Fig. 11.6 Relationship between SDGs and sustainable manufacturing strategies

two of the largest and most important companies in the textile sector in Colombia and, at the end of 2002, they merged to form a single company called Fabricato-Tejicondor Textiles.

In terms of sustainable manufacturing strategies, it has been found that they have focused on the implementation of more efficient production processes that reduce energy consumption and waste generation. In addition, they have started to use recycled materials in their products. They have also implemented cleaner and more efficient production technologies to reduce carbon emissions and minimise their environmental impact.

Other textile industry who implemented sustainable strategies is Coltejer S.A.: Compañía Colombiana de Tejidos—Coltejer S.A., started operations in 1907, but entered a process of suspension of operations in 2021. Although it is not an example of corporate sustainability due to this situation, it is important to highlight that, during its time of operations, this Colombian company implemented measures to reduce water consumption in its textile production processes, adopting water recycling technologies, reusing water in industrial processes, optimising purification, and effluent treatment systems to minimise its water footprint and the contamination of water bodies near its production plants.

In addition to its focus on water management, the company is committed to innovation in materials and processes to promote sustainability in its textile products, incorporating recycled fibres and organic materials in its fabrics, thereby reducing dependence on non-renewable natural resources, and contributing to the circular economy. Another fundamental aspect of the sustainability strategy is its commitment to the social dimension through the implementation of welfare programmes for its employees, promoting safe and dignified working conditions in all its facilities and working closely with local communities to promote socio-economic development and improve the quality of life of those living in the areas where it operates.

Another industry in this selected group is Enka de Colombia: This textile company was founded in 1964 by the Dutch group Akzo Nobel with the purpose of manufacturing and marketing resins and synthetic fibres, becoming the leading manufacturer of nylon 6 tyre canvas in the Americas and the largest producer of filaments and synthetic fibres in the Andean Region. This company has developed a circular economy strategy based on a sustainable supply chain that allowed it, in 2023, sales of close to 31,500 tonnes of recycled resin, capturing more than 1,354 million post-consumer PET bottles, contributing to the social and economic development of more than 100,000 recyclers in the country, thus becoming one of the largest and most modern PET recycling plants in the world. Moreover, as part of the operational optimisation and sustainability strategy, Enka increased the consumption of biomass in energy generation, which enabled a reduction equivalent to 22,600 tonnes of CO2, having a favourable impact on environmental aspects and diversification of fuel supply.

Other important case is Fibretex: Fibras y Recuperación Textil (Fibretex) is a Colombian company founded in the mid 70's of the last centuries, which has revolutionised the textile sector by developing an innovative process of textile fibre production from textile surplus (remnants) and solid waste. This circular economy approach, pioneered in the textile sector, not only contributes to the reduction of solid waste, but also offers a sustainable alternative to conventional fibres such as cotton and polyester.

The textile fibre production process begins with the collection of solid waste, which can include plastics, paper, cardboard, and unwanted textiles. These materials undergo a rigorous sorting and separation process to remove contaminants and other materials that cannot be recovered in the process. The waste is then shredded and undergoes a chemical and mechanical transformation process to convert it into usable textile fibres. One of the key advantages of the Fibretex process is its versatility as the company can produce a wide range of textile fibres with different properties and characteristics, tailored to the specific needs of its customers in the textile industry. These fibres can be used in the manufacture of a variety of products, including apparel, accessories, household products and industrial materials. Fibretex's impact on the textile industry goes beyond the production of sustainable fibres. The company is also contributing to changing the perception of waste as a problem to see it as an opportunity.

By turning waste into valuable resources, the company is demonstrating the potential of the circular economy to transform industries and create a more sustainable future. In addition to its positive environmental impact, the production of textile fibre from recovered textile waste and solid waste also has significant social benefits. Job creation in waste collection and processing can help reduce poverty and improve living conditions in vulnerable communities.

And finally, other textile industry in this sustainable group is Vélez: Cueros Vélez, S.A.S., a well-known Colombian fashion brand founded in 1986 and specialising in the leather sector, has stood out for its commitment to sustainable manufacturing, implementing various strategies to reduce its environmental impact and promote responsible practices throughout its value chain. One of Velez's main sustainable manufacturing strategies focuses on the selection of materials and production processes that minimise environmental impact. The company is committed to using organic and sustainably sourced materials in the manufacture of its textile products. This includes the use of leather from certified sources and the use of organic cotton in the manufacture of garments.

By opting for materials that meet rigorous environmental and social standards. In addition to the careful selection of materials, Velez has implemented measures to reduce the waste of resources during the production process. The company has adopted technologies and practices that optimise the use of materials and minimise waste generation. This includes the efficient design of patterns to maximise the use of raw materials and the implementation of manufacturing processes that reduce water and energy consumption. Velez has also developed programmes for recycling, reusing materials and refurbishing garments and accessories to give products a second life at the end of their useful life cycle. Another key aspect of Velez's sustainable manufacturing strategy is its commitment to transparency and traceability throughout its supply chain. The company works closely with its suppliers to ensure that they meet high ethical and environmental standards in the production of raw materials and components.

Through regular audits and assessments, Velez verifies compliance with criteria such as respect for labour rights, environmental protection, and animal welfare. In addition, the company has implemented programmes such as training and professional development initiatives for staff, as well as social investment projects in areas such as education, health, and community development. In addition, Velez collaborates with non-governmental organisations and local entities to support economic and social development initiatives in the most vulnerable communities.

These companies represent just a sample of the Colombian textile sector's growing commitment to sustainable manufacturing. Now then, expanding the scope of phase 2 a bit further and paving the way for phase 3, a projection is made of the challenges and perspectives that the Colombian textile manufacturing sector faces in the horizon.

Regarding the prospects and challenges faced by the textile sector in terms of sustainability, the following can be highlighted: Sustainability in textile manufacturing involves adopting practices that minimize environmental impact and promote resource efficiency. This includes reducing carbon emissions, using recycled and sustainable materials, efficient water, and energy management, and implementing cleaner and safer production processes.

One of the main challenges faced by the Colombian textile sector is international competition, especially from Asian countries where production costs are lower. However, adopting sustainable practices can become a competitive advantage. Consumers, both domestic and international, are increasingly aware of the importance of sustainability and are willing to pay more for products that align with these values. Thus, Colombian textile companies can differentiate themselves in the global market by offering high-quality and sustainable products.

Innovation also plays a crucial role in the transition towards sustainable manufacturing. Textile companies in Colombia are investing in advanced technologies that enable more efficient use of resources. For example, waterless dyeing technology and textile recycling processes are key areas of development. Additionally, the use of renewable energy is being explored to reduce dependency on fossil fuels.

Discussion and Conclusions

This chapter develops the third phase of the study, according to the methodology, and presents the critical evaluation of successful strategies of sustainable manufacturing applied for the textile industries analysed. Colombia has shown a strong commitment to sustainability in the textile sector. Several companies have adopted eco-friendly practices, and the government has implemented policies to encourage sustainable manufacturing.

After studying the cases presented above, it can be affirmed that, although the textile manufacturing sector is, as Xu et al. (2018) say, one of the most unsustainable in environmental, social and economic terms, part of the Colombian textile industrial sector has made sustainability an increasingly relevant issue in a world concerned about the environmental and social impact of garment production, leading the country to see an increase in awareness of sustainable practices in this industry in recent years.

This study also allowed us to validate Caicedo Manrique's (2021) assertions regarding the existence of limitations related to investment in technology, innovation, and growth. In this regard, it was found that, despite being a generality in the industries of the sector, there is a plural number of textile industries that have developed processes of technological updating, innovation and competitiveness through the adoption of sustainable manufacturing strategies that have allowed them to generate differential quality and strategic positioning, obtaining a competitive advantage in the market, reaffirming what was said by Chang and Cheng (2019); Bhupendraa and Sangleb (2016); Lukena et al. (2016) and Severo et al. (2017).

It also allowed contrasting the assertions of Salgado Abril and Suárez Cordero (2016), confirming the social impact, in terms of employment, of the textile sector in the country. And those of Mazo et al. (2020), by showing that, although there are many companies in this sector in our country that manage precarious working conditions on some occasions, with low wages, long working hours and lack of job security. There are some companies that, in search of social sustainability, have implemented sustainable manufacturing strategies that address these conditions and generate recognition in society.

Regarding the sustainable manufacturing strategies implemented by industries in the Colombian textile manufacturing sector and their impact on the SDGs, this study validated the assertions of Desore and Narula (2018) on the role of manufacturing companies in solving problems related to sustainable development, using creativity and innovation, transcending the economic dimension of sustainable development by aiming for a harmonious polyhedron as presented by Chica and Serna (2018).

With regard to SDG 12 and its relationship with the Colombian textile industry, we contrasted what was proposed by Ayala Rodríguez (2021); Bick et al. (2018) and it was found that, although their statements are true, some of the companies in the Colombian textile sector have addressed this problem by adopting sustainable manufacturing practices, reducing resource consumption, minimising waste and promoting the reuse and recycling of materials, generating interesting results worthy of replication in other companies in this sector that have not yet decisively addressed these strategies, as presented by Niinimäki et al. (2020).

Similarly, when addressing SDG 8, it was found that it has a direct connection to the textile industry. The direct and indirect employment impact of garment production is estimated at 1,000,000 people, confirming what was stated by de Oliveira Neto et al. (2019). However, with respect to the statement presented by Tabales et al. (2021) and Herrera Almanza and Corona (2020); it was possible to contrast, in the successful experiences highlighted in this research, that there are companies that, through the implementation of sustainable manufacturing strategies from a harmonic polyhedron approach, are facing up to these statements and generating a change in the social and labour aspects of the industries.

Continuing with the study of the impact of manufacturing strategies on SDG 9, it was found that the adoption of clean and efficient technologies in textile production is fundamental to reduce its negative impact on the environment, as stated by Edreira Rico (2022). Furthermore, improving infrastructure in the areas where textile factories operate, promoting research, innovation and development in sustainable textiles and fostering collaboration between companies, governments and civil society are essential steps towards a more equitable, innovative, and sustainable textile sector, as stated by Herrera Almanza and Corona (2020) and Kynčlová et al. (2020).

In another relevant goal for the textile industry, SDG 6, although it was found that the production of garments requires large amounts of water, especially in dyeing and finishing processes, the successful cases of the Colombian textile industry addressed in this research, through the adoption of more efficient technologies and production practices in the use of water, reducing pollution from industrial discharges and working on the conservation and sustainable management of water resources, managed to generate positive impacts on the reduction of their water footprint and on SDG6.

Another finding of this study is that the textile industry is also linked to other SDGs indirectly. For example, SDG 1: End poverty, relates to the textile industry through its impact on employment and livelihoods of local communities. Similarly, SDG 13: Climate action, is affected by the greenhouse gas emissions and energy consumption associated with the production and transport of garments.

Thus, as a result of the critical evaluation of this study, it can be concluded that the sustainable manufacturing strategies implemented by the case studies of textile manufacturing companies analysed in this research have had a positive impact on the achievement of outcomes that impact the targets of the Sustainable Development Goals, in particular the targets corresponding to SDGs 6, 8, 9 and 12, and more broadly SDGs 1 and 13. Thus confirming that the implementation of sustainable manufacturing is an efficient and effective alternative to gain a competitive advantage while achieving business objectives and is fully aligned with the SDG targets.

Education and training are fundamental to support this transition. Universities and research centres in Colombia are collaborating with the textile industry to develop training programs and research projects that address sustainability challenges. This not only enhances the technical capacity of the sector but also fosters a culture of innovation and environmental responsibility. Moreover, strategic partnerships between the public and private sectors are crucial to advancing sustainable manufacturing. Collaborations between textile companies, material suppliers, non-governmental organizations, and government entities can drive the development of innovative and sustainable solutions. These alliances allow for the sharing of knowledge, resources, and technologies, accelerating progress towards a more sustainable textile industry.

However, it should be noted that there are still challenges to be faced, such as the proper management of textile waste and the reduction of overconsumption. Consumer education is also crucial to promote the purchase of sustainable clothing and foster a conscious fashion culture. The textile sector in Colombia has significant potential to lead in the realm of sustainable manufacturing. Although it faces challenges, the opportunities are immense. The combination of a strong commitment to sustainability, investment in technological innovation, education and training, and government support creates a conducive environment for the sustainable growth of the textile industry. With a well-defined strategy and effective collaboration, Colombia can position itself as a leader in sustainable textile manufacturing globally, contributing not only to the country's economic development but also to environmental protection and social well-being.

References

Abbate S, Centobelli P, Cerchione R, Nadeem SP, Riccio E (2024) Sustainability trends and gaps in the textile, apparel and fashion industries. Environ Dev Sustain 26(2):2837–2864

Accinelli Gamba E, De la FuenteGarcía JL (2013) Responsabilidad social corporativa, actividades empresariales y desarrollo sustentable Modelo matemático de las decisiones en la empresa. Contaduría y Administración 58(3):227–248. https://doi.org/10.1016/S0186-1042(13)71228-2

Agrawal S, Singh RK (2019) Analyzing disposition decisions for sustainable reverse logistics: triple bottom line approach. Res Conserv Recycl 150:104448

Akter MMK, Haq UN, Islam MM, Uddin MA (2022) Textile-apparel manufacturing and material waste management in the circular economy: A conceptual model to achieve sustainable development goal (SDG) 12 for Bangladesh. Clean Environ Syst 4:100070

Andrade-Pérez GI, Avella Rodríguez C, Baptiste-Ballera BL, Bustamante Zamudio C, Chaves ME, Corzo G, Trujillo M (2018) Transiciones socioecológicas hacia la sostenibilidad: gestión de la biodiversidad en los procesos de cambio de uso de la tierra en el territorio colombiano. Instituto de Investigación de Recursos Biológicos Alexander von Humboldt

Antonucci CF (2021) La problemática de la obsolescencia programada en los bienes de consumo

Argüello Gavilánez AE (2022) Análisis del ciclo de vida, consumo y recuperación de indumentaria para el desarrollo de colecciones sustentables (Bachelor's thesis, Pontificia Universidad Católica del Ecuador)

Assembly G (2015) Sustainable development goals. SDGs transform our world, 2030(10.1186)
Baah C, Opoku-Agyeman D, Acquah ISK, Agyabeng-Mensah Y, Afum E, Faibil D, Abdoulaye FAM (2021) Examining the correlations between stakeholder pressures, green production practices, firm reputation, environmental and financial performance: Evidence from manufacturing SMEs. Sustain Prod Consum 27:100–114
Babativa Chirivi SM, Babilonia Ballesteros RI, Pérez D (2023) Braiding the identity of the zenú people: territory and nature-culture relationships in the crafting of the Vueltiao Hat. TEXTILE, 1–17
Baliyan R, Chaudhary MD, Verma SK, Girotra R, Madhukar KS (2024) A holistic approach towards textile industry in post-pandemic era. J Res Adm 6(1)
Baumgartner RJ, Ebner D (2010) Corporate sustainability strategies: sustainability profiles and maturity levels. Sustain Dev 18(2):76–89
Bick R, Halsey E, Ekenga CC (2018) The global environmental injustice of fast fashion. Environ Health 17:1–4
Boschmeier E, Ipsmiller W, Bartl A (2024) Market assessment to improve fibre recycling within the EU textile sector. Waste Manag Res 42(2):135–145
Bradu P, Biswas A, Nair C, Sreevalsakumar S, Patil M, Kannampuzha S, Gopalakrishnan AV (2023) Recent advances in green technology and Industrial Revolution 4.0 for a sustainable future. Environ Sci Pollut Res 30(60):124488–124519
Brundtland GH (1987) Our common future: report of the world commission on environment and development. United Nations Commission 4(1):300. https://doi.org/10.1080/07488008808408783
Bustos Mundaca MI (2022) Experiencias de emprendimiento en moda lenta: propuestas alternativas transitando entre racionalidades
Cai YJ, Choi TM (2020) A United Nations' sustainable development goals perspective for sustainable textile and apparel supply chain management. Transp Res Part E: Logist Transp Rev 141:102010
Caicedo Manrique CE (2021) La realidad del sector textil colombiano
Camilleri MA (2022) Strategic attributions of corporate social responsibility and environmental management: The business case for doing well by doing good! Sustain Dev 30(3):409–422
Carrera Gallissà E (2017) Los retos sostenibilistas del sector textil. Rev Quím E Ind Text 220:20–32
Castell OC, Pérez RMO, Murillo LFS, Ospina NAR, Núñez KVR (2016) Análisis de las políticas de comercio exterior en Colombia para disminuir el ingreso de mercancías del sector textil y calzado provenientes de China. Revista Vía Iuris 21:127–146
Castro MFG (2021) La gestión de talento humano y su influencia en el desempeño laboral para el éxito de las empresas. Polo del Conocimiento: Revista científico-profesional 6(8):318–329
Chang A-Y, Cheng Y-T (2019) Analysis model of the sustainability development of manufacturing small and medium- sized enterprises in Taiwan. J Clean Prod 207:458–473. https://doi.org/10.1016/j.jclepro.2018.10.025
Cheng Z, Sun Z, Zhu S, Lou Z, Zhu N, Feng L (2019) The identification and health risk assessment of odor emissions from waste landfilling and composting. Sci Total Environ 649:1038–1044
Chica-Urzola JA, Serna Mendoza C (2018) A new harmonic polyhedral structure based in life and well-being as framework to solve the main problem of the sustainable development. Eur J Sustain Dev 7(4):391–401. https://doi.org/10.14207/ejsd.2018.v7n4p391
Choi M (2023) Impactos de la implementación de la moda rápida en las condiciones laborales de las trabajadoras de la industria de la confección en Bangladesh
Ciplet D (2021) From energy privilege to energy justice: a framework for embedded sustainable development. Energy Res Soc Sci 75:101996
Clemente-Suárez VJ, Rodriguez-Besteiro S, Cabello-Eras JJ, Bustamante-Sanchez A, Navarro-Jiménez E, Donoso-Gonzalez M, Tornero-Aguilera JF (2022) Sustainable development goals in the COVID-19 pandemic: a narrative review. Sustainability 14(13):7726

Correa Morales KV (2023) Customización de vestuario basada en técnicas de upcycling e intervención textil como estrategias de economía circular para la reducción de residuos textiles en la industria de la moda
Cortés Hernández A, Tinoco VB, García Torres AB (2023) Responsabilidad social empresarial en la industria textil del fast fashion: preocupación o simulación?. Revista FACE 23(3)
Cubero MC (2021) La Responsabilidad Social Empresarial y el Greenwashing como estrategias empresariales: efectos sobre el medio ambiente. Caso de la industria textil. Obs Medioambient 24
da Silva PC, de Oliveira Neto GC, Correia JMF, Tucci HNP (2021) Evaluation of economic, environmental and operational performance of the adoption of cleaner production: survey in large textile industries. J Clean Prod 278:123855
de Oliveira Neto GC, Correia JMF, Silva PC, de Oliveira Sanches AG, Lucato WC (2019) Cleaner Production in the textile industry and its relationship to sustainable development goals. J Clean Prod 228:1514–1525
De Sociedades S (2017) Desempeño del sector textil-confección. Informe. Recuperado de https://incp.org.co/Site/publicaciones/info/archivos/Textiles.pdf.
Desore A, Narula SA (2018) An overview on corporate response towards sustainability issues in textile industry. Environ Dev Sustain 20:1439–1459
Edreira Rico L (2022) Contribución a la mejora de la competitividad en la industria textil: ODS
Escobar AAH, Rodríguez MPR, López BMP, Ganchozo BI, Gómez AJQ, Ponce LAM (2018) Metodología de la investigación científica (Vol. 15). 3Ciencias
Fayomi GU, Mini SE, Fayomi OSI, Ayoola AA (2019) Perspectives on environmental CO2 emission and energy factor in Cement Industry. In: IOP conference series: earth and environmental science, vol. 331, No. 1, p. 012035. IOP Publishing
Fernández-López S, Rodeiro-Pazos D, Rey-Ares L (2020) Effects of working capital management on firms' profitability: evidence from cheese-producing companies. Agribusiness 36(4):770–791
Gabriel M, Luque MLD (2020) Sustainable development goal 12 and its relationship with the textile industry. The UN Sustainable Development Goals for the Textile and Fashion Industry, 21–46
Garza EG (2007) De las teorías del desarrollo al desarrollo sustentable. Historia de la construcción de un enfoque multidisciplinario. Trayectorias 9(25):45–60
Gbededo MA, Liyanage K, y Garza-Reyes, J. A. (2018) Towards a life cycle sustainability analysis: a systematic review of approaches to sustainable manufacturing. J Clean Prod 184:1002–1015. https://doi.org/10.1016/j.jclepro.2018.02.310
Gbolarumi FT, Wong KY, Olohunde ST (2021) Sustainability assessment in the textile and apparel industry: a review of recent studies. In IOP conference series: materials science and engineering, vol. 1051, No. 1, p. 012099. IOP Publishing
Gómez ÓT, Guevara LR, López PR (2009) Perspectivas de la moda sostenible en el Perú. Industrial data 12(2):68–72
González PAE, Soto DMA, Ramírez ÁJM (2018) Sector textil colombiano y su influencia en la economía del país. Punto de vista 9(1):4
Granados MLP, Chavez MTA (2021) Fast Fashion:¿ moda o contaminación? Jóvenes En La Ciencia 12:1–8
Harsanto B, Primiana I, Sarasi V, Satyakti Y (2023) Sustainability innovation in the textile industry: a systematic review. Sustainability 15(2):1549
Henao R, Sarache W, Gómez I (2019) Lean manufacturing and sustainable performance: trends and future challenges. J Clean Prod 208:99–116. https://doi.org/10.1016/j.jclepro.2018.10.116
Herrera Almanza AM, Corona B (2020) Using social life cycle assessment to analyze the contribution of products to the sustainable development goals: a case study in the textile sector. Int J Life Cycle Assess 25:1833–1845
Hu Y, Cheng H, Tao S (2017) Environmental and human health challenges of industrial livestock and poultry farming in China and their mitigation. Environ Int 107:111–130

Instituto para la exportación y la moda (Inexmoda). Observatorio de moda febrero 2024. Retrieved in: https://www.inexmoda.org.co/wp-content/uploads/2024/02/Observatorio-Inexmoda-Febrero-2024-F.pdf (03/15/2024)

Islam MT, Jahan R, Jahan M, Howlader MS, Islam R, Islam MM, Robin AH (2022) Sustainable textile industry: an overview. Non-Metallic Mater Sci 4(2):15–32. https://doi.org/10.30564/nmms.v4i2.4707

Jarpa SG, Halog A (2021) Pursuing a circular and sustainable textile industry in Latin America. Sustain Fash TextEs Lat Am 105–130

Jayaraman K, Jayashree S, Dorasamy M (2023) The effects of green innovations in organizations: influence of stakeholders. Sustainability 15(2):1133

Jiang Q, Liu Z, Liu W, Li T, Cong W, Zhang H, y Shi, J. (2018) A principal component analysis based three-dimensional sustainability assessment model to evaluate corporate sustainable performance. J Clean Prod 187:625–637. https://doi.org/10.1016/j.jclepro.2018.03.255

Jiménez Castellón S (2023) Análisis del mercado textil actual, tendencia Fast Fashion y el consumo masivo, economía lineal o economía circular

Kynčlová P, Upadhyaya S, Nice T (2020) Composite index as a measure on achieving sustainable development goal 9 (SDG-9) industry-related targets: The SDG-9 index. Appl Energy 265:114755

Leal Filho W, Azeiteiro U, Alves F, Pace P, Mifsud M, Brandli L, Disterheft A (2018) Reinvigorating the sustainable development research agenda: the role of the sustainable development goals (SDG). Int J Sustain Dev World Ecol 25(2):131–142

Linares EL, Galeano G, García N, Figueroa Y (2008) Fibras vegetales empleadas en artesanías en Colombia

Lukena RA, Van Berkelb R, Leuenbergerc H, Schwagerb P (2016) A 20-year retrospective of the national cleaner production centres programme. J Clean Prod 112: 1165–1174

Maldonado-Pinto JE, Portilla-Barco LF (2020) Procesos de innovación en la industria manufacturera colombiana (Innovation Processes in the Colombian Manufacturing Industry). Revista CEA, 6(11)

Martínez LIV (2021) Cadenas de suministro manchadas: Explotación laboral en la industria de la moda de China e India. Online Journal Mundo Asia Pacifico 10(19):85–97

Mazo AZ, Arenas JAC, Peláez MM (2018) Gestión logística en el sector textil-confección en Colombia: retos y oportunidades de mejora para la competitividad. Clío América 12(23):98–108

Mazo AJ, Montoya Cano MDM, Pulgarín Rondón AS (2020) Retos y oportunidades de la industria textil y confección en el momento actual en Colombia

Meadows DHM (1972) The limits to growth. The Club of Rome. https://doi.org/10.1111/j.1752-1688.1972.tb05230.x

Medina AIT, Escalante RYS, Prada RAP, Boada CAG (2019) Tendencias en la productividad del sector textil-confecciones en Colombia. Visión Internacional (Cúcuta), 12–16

Medina Fábrega CM (2023) La comunicación en moda sostenible y su eficacia-Medina Fábrega, Carmen

Moldavska A, y Welo, T. (2017) The concept of sustainable manufacturing and its definitions: a content-analysis based literature review. J Clean Prod 166:744–755. https://doi.org/10.1016/j.jclepro.2017.08.006

Morioka SN, de Carvalho MM (2016) A systematic literature review towards a conceptual framework for integrating sustainability performance into business. J Clean Prod 136:134–146. https://doi.org/10.1016/j.jclepro.2016.01.104

Nguyen HT, Le DMD, Ho TTM, Nguyen PM (2021) Enhancing sustainability in the contemporary model of CSR: a case of fast fashion industry in developing countries. Soc Responsib J 17(4):578–591

Nguyen GN, Mani V, Kha MK, Papadopoulos T (2023) Supply chain social responsibility in labour-intensive industries: a practitioner's perspective. Prod Plan Control 34(4):371–390

Niinimäki K, Peters G, Dahlbo H, Perry P, Rissanen T, Gwilt A (2020) The environmental price of fast fashion. Nat Rev Earth Environ 1(4):189–200

Norando V (2020) Mujeres y niñas entre hilos: condiciones de trabajo en la industria textil a principios del siglo XX. Una experiencia de clase generizada. Mora (Buenos Aires) 26(1):51–60
Pagell M, Shevchenko A (2014) Why research in sustainable supply chain management should have no future. J Supply Chain Manag 50(1):44–55
Papamichael I, Chatziparaskeva G, Pedreño JN, Voukkali I, Candel MBA, Zorpas AA (2022) Building a new mind set in tomorrow fashion development through circular strategy models in the framework of waste management. Curr Opin Green Sustain Chem 36:100638
Pastor Tomás L (2023) RSC, sostenibilidad y consumo responsable en la industria de la moda
Pinilla JMC, González AB (2014) Estrategias de marketing sectorial. ESIC Editorial
Pitre-Redondo R, De la Ossa-Guerra S, Hernández-Palma H (2020) Ecoeficiencia: clave de la responsabilidad ambiental empresarial en el sector textil. Desarrollo Gerencial 12(2):1–20
Pradhan P, Costa L, Rybski D, Lucht W, Kropp JP (2017) A systematic study of sustainable development goal (SDG) interactions. Earth's Future 5(11):1169–1179
Quintero Arango LF, Álvarez Agudelo CE, Ibagon Parra DM (2019) La violencia intrafamiliar en el desempeño laboral de las mujeres en el sector textil-confección. Diversitas: perspectivas en psicología 15(2): 271–284
Rathore B (2022). Textile Industry 4.0 transformation for sustainable development: prediction in manufacturing & proposed hybrid sustainable practices. Eduzone Int Peer Rev/Refereed Multidiscip J 11(1):223–241
Roberts SJF, Ball PD (2014) Developing a library of sustainable manufacturing practices. Proc CIRP 15:159–164. https://doi.org/10.1016/j.procir.2014.06.054
Rodríguez Bugada JP (2022) Pe-er: la revolución de la moda sustentable
Rosen MA, Kishawy HA (2012) Sustainable manufacturing and design: concepts, practices and needs. Sustainability 4(2):154–174
Saha K, Dey PK, Papagiannaki E (2022) Implementing circular economy in the textile and clothing industry. In: Supply chain sustainability in small and medium sized enterprises, pp 239–276. Routledge
Salgado Abril SD, Suárez Cordero JA (2016) Análisis de la competitividad de las empresas del sector textil en Colombia desde un enfoque sistémico (2008–2013)
Sanchez-Planelles J, Segarra-Oña M, Peiro-Signes A (2022) Identifying different sustainable practices to help companies to contribute to the sustainable development: Holistic sustainability, sustainable business and operations models. Corp Soc Responsib Environ Manag 29(4):904–917
Santiago M (1982) La industria textil en Colombia: 1900–1945. Revista Desarrollo y Sociedad
Segurajáuregui Álvarez L (2016) Las diversas caras de la obsolescencia
Sharma A, Syed Z, Brighu U, Gupta AB, Ram C (2019) Adsorption of textile wastewater on alkali-activated sand. J Clean Prod 220: 23–32
Sheffi Y (2023) The magic conveyor belt: supply chains, AI, and the future of work. MIT CTL Media
Silva S, Lopes AM (2017) Environmental aspects and impacts of a waste incineration plant. Energy Proc 136:239–244
Silván-Ferrero P, Silván-Martínez P, Trespalacios JA, Sedano IF (2023) Sostenibilidad y economía circular en el sector textil: análisis del consumo de ropa de segunda mano como alternativa de moda sostenible.: Sustentabilidad y economía circular en el sector textil. Behanomics, 1
Tabales JMN, del Amor Collado E, Carmona FJR (2021) Economía circular en la industria de la moda: pilares básicos del modelo. Revista de ciencias sociales 27(4):162–176
Thakker AM, Sun D (2023) Sustainable development goals for textiles and fashion. Environ Sci Pollut Res 30(46):101989–102009
UNDP (United Nations Development Programme) (2016) Human Development Report 2016: Human Development for Everyone. New York
Vildåsen SS, Keitsch MY, Fet AM (2017) Clarifying the epistemology of corporate sustainability. Ecol Econom 138:40–46. https://doi.org/10.1016/j.ecolecon.2017.03.029

Vinlove AL (2021) Ropas que importan. Utilizando el Upcycling como intervención en el sistema de la Moda e Indumentaria para transicionar hacia futuros sostenibles. Cuadernos del Centro de Estudios de Diseño y Comunicación (132)

Xu CK, Cheng H, Liao ZJ (2018) Towards sustainable growth in the textile industry: a case study of environmental policy in China. Polish J Environ Stud 27(5):2325–2336. https://doi.org/10.15244/pjoes/79720

Yadav R, Kamble Z (2024) Textile waste-based cellulose composites: a review. J Mater Sci 1–22

Zaman SI, Kusi-Sarpong S (2024) Identifying and exploring the relationship among the critical success factors of sustainability toward consumer behavior. J Model Manage 19(2):492–522

Conclusions

Sustainability in the textile industry is a multifaceted issue that requires a holistic, multidisciplinary approach to address the many environmental, social, economic challenges facing the sector. Throughout this book, we have explored various strategies, practices that are being implemented in different regions of the Americas to promote sustainability in the textile industry. Not only are these initiatives crucial to meeting the SDGs by 2030, but they are also essential to ensure the long-term viability of the sector

Section I has demonstrated the importance of sustainable practices in the textile value chain, highlighting how the integration of advanced techniques such as geospatial analysis can identify critical areas for emission reduction and environmental protection. Chapter 1 provides a solid foundation for effective policymaking and highlights the need for data-driven approaches to address environmental challenges, aligning with SDGs 9 (Industry, Innovation, and Infrastructure) and 12 (Responsible Consumption and Production). Chapter 2 also highlights the technological advances that are driving sustainability in the sector, contributing to the SDGs themselves.

Eco-innovation is presented as a vital component of sustainability in the textile industry in Section II. The case studies in Colombian exporting companies, presented in Chap. 3 demonstrate that the adoption of eco-innovation practices not only reduces environmental impact but also improves competitiveness and organizational performance, aligning with SDGs 8 (Decent Work and Economic Growth) and 13 (Climate Action). This approach is critical in an environment where consumer expectations and government regulations are constantly evolving towards greener standards.

Transformational and sustainable leadership, explored in Chap. 4, is a critical factor influencing the success of sustainable practices. The findings of the study in Peru highlight how leadership focused on individualized support, inspirational motivation, and intellectual stimulation can foster an engaged and sustainability-oriented work environment. This type of leadership not only improves employee well-being but also aligns individual goals with organizational sustainability goals, connecting with SDGs 8 and 12.

A. E. Toscano-Hernández et al. (eds.), *Sustainability, Stakeholders and Marketing in the Textile Sector*, SDGs and Textiles, https://doi.org/10.1007/978-981-97-6952-0

The barriers to innovation in textile SMEs, as discussed in Chap. 5, underline the need for a more inclusive and supportive approach in terms of knowledge and financial resources. Identifying these barriers provides a basis for designing supportive policies and programs that facilitate innovation and the adoption of sustainable practices in this crucial segment of the industry, aligning with SDGs 9 and 8.

Reverse logistics and the circular economy, particularly in the Mexican context, emerge as fundamental tools for sustainability. The effective implementation of reverse logistics programs, as discussed in Chap. 6 can close the life cycle of textile products, reducing waste and promoting reuse and recycling, contributing to SDGs 11 and 12. However, the success of these programs depends to a large extent on the effectiveness of internal processes and business policies.

Section III highlights green marketing as a powerful strategy for engaging consumers in sustainability. Case studies of sustainable fashion companies in Colombia, presented in Chap. 7 show how green marketing strategies, including consumer education, environmental labeling, and product innovation, can improve consumer perception and engagement with sustainable practices. These strategies not only help position companies as environmentally responsible but also encourage greater demand for sustainable products, aligning with SDGs 12 and 10.

Finally, the green swan theory and its application to the circular economy and regenerative capitalism, discussed in Chap. 8 offer a critical and forward-looking perspective on the transformations needed in the global economic system to effectively address the challenges of climate change and sustainability. The adoption of economic models that integrate principles of regeneration and circularity can provide a viable pathway to a more sustainable and equitable future, connecting with SDGs 9 and 13.

In conclusion, this book has highlighted the importance of taking a comprehensive and collaborative approach to promoting sustainability in the textile industry. The strategies and practices presented not only offer specific solutions to current challenges but also provide a framework for future research and action. By integrating diverse perspectives and innovative approaches, we hope that this work will inspire academics, practitioners, and policymakers to continue working toward a more sustainable and resilient textile sector.